International Political Economy Series

Series Editor
Timothy M. Shaw
University of Massachusetts Boston
Boston, MA, USA

Emeritus Professor
University of London
London, UK

The global political economy is in flux as a series of cumulative crises impacts its organization and governance. The IPE series has tracked its development in both analysis and structure over the last three decades. It has always had a concentration on the global South. Now the South increasingly challenges the North as the centre of development, also reflected in a growing number of submissions and publications on indebted Eurozone economies in Southern Europe. An indispensable resource for scholars and researchers, the series examines a variety of capitalisms and connections by focusing on emerging economies, companies and sectors, debates and policies. It informs diverse policy communities as the established trans-Atlantic North declines and 'the rest', especially the BRICS, rise. NOW INDEXED ON SCOPUS!

More information about this series at
http://www.palgrave.com/gp/series/13996

Jean-François Rousseau ·
Sabrina Habich-Sobiegalla
Editors

The Political Economy of Hydropower in Southwest China and Beyond

palgrave
macmillan

Editors
Jean-François Rousseau
School of International Development
and Global Studies
University of Ottawa
Ottawa, ON, Canada

Sabrina Habich-Sobiegalla
Institute of Chinese Studies
Freie Universität Berlin
Berlin, Germany

ISSN 2662-2483 ISSN 2662-2491 (electronic)
International Political Economy Series
ISBN 978-3-030-59360-5 ISBN 978-3-030-59361-2 (eBook)
https://doi.org/10.1007/978-3-030-59361-2

Cover credit: RBFried/iStockphoto

This Palgrave Macmillan imprint is published by the registered company Springer Nature Switzerland AG
The registered company address is: Gewerbestrasse 11, 6330 Cham, Switzerland

Acknowledgements

This book exists because of the warm enthusiasm that some people have expressed towards our project since its onset. These individuals deserve our recognition and we wish to thank them for their precious help and support.

The first among these persons is Timothy Shaw, an individual who wears many hats, including that of founding editor-in-chief of the IPE series. Since we first discussed the idea of putting this collection together a few years ago, Tim has always been supportive. He has helped us in navigating the procedures we had to go through for this project to become a reality and shared tonnes of information and references that contributed to making this a better book.

We also wish to thank the contributors to this volume, who have kindly agreed to share their knowledge and insight in this outlet. The authors whose work is published herein have been exceptional at dealing with our multiple deadlines and email queries and generously contributed original and significant findings.

Thomas Kettig is also a person that wears many hats, including those of linguist and copy-proofer. His work is present throughout this manuscript (even in these acknowledgements!) and has enhanced the quality of this volume in countless ways. We also thank Qiang Li and Mitsy Barriga Ramos for editorial assistance.

At Palgrave Macmillan, Anca Pusca has been a very encouraging editor whose support has been key at every step along the way. Rachel Moore also kindly helped us in the initial stages of this project, while Arun Kumar Anbalagan and Preetha Kuttiappan did so during the final ones. We also thank two anonymous for their comments and suggestions.

The editors acknowledge funding from the Social Sciences and Humanities Research Council of Canada, University of Ottawa Faculty of Social Sciences (Research Group Program) and the German Research Foundation (DFG) (Grant 401070338).

Contents

Notes on Contributors

Ricardo Andrade is doctoral researcher at the Otto Suhr Institute of Political Science, Freie Universität Berlin. His research interests include Brazil-China relations, hydropower, socio-technical and socio-environmental systems, public participation, algorithmic governance, and democracy. Trained in Communications, Public Management and Technology, his professional experience includes management positions at NGOs and in the private sector in Brazil, Europe, and China. He has participated in EU-funded research projects on innovation and public participation.

Sebastian Biba is a research fellow at the Institute of Political Science at Goethe University Frankfurt, Germany. His research interests lie in the field of China's foreign policy, including a focus on China's international river politics. He is the author of the monograph *China's Hydro-politics in the Mekong* and his articles have appeared in journals such as *Security Dialogue*, *Third World Quarterly*, *Journal of Contemporary China*, and *Water International*.

Zhuo Chen is a graduate student in Applied Anthropology at Oregon State University. She holds a B.S. in Hydrology and Water Resources Engineering from Wuhan University, China. Her research interests focus on the social impacts of science and technology.

Yuefang Duan is a professor of Resettlement Economics at China Three Gorges University in the School of Economics and Management, and the Director of the Research Center for Reservoir Resettlement, which is the key research base of humanities and social sciences in Hubei Province, China. His research interests mainly focus on involuntary resettlement that is caused by development projects. He has been involved in numerous research projects and published more than 80 papers on Chinese resettlement policies and practice.

Bettina Gransow is Associate Professor at the Institute of Chinese Studies, Freie Universität Berlin. Her research interests include social risk analysis of Chinese infrastructure projects; migration, megacities and urban redevelopment in China from a sociological perspective. She recently published articles and book chapters related to Chinese infrastructure investments in Latin America, social risk management at AIIB (together with Susanna Price), urban sociology and contemporary urbanism in China, migrants in the Pearl River Delta and urban redevelopment in Guangzhou.

Sabrina Habich-Sobiegalla is Assistant Professor at the Institute of China Studies at Freie Universität Berlin. Her research interests include regional development, central-local relations, and energy and resource governance with a focus on China. She is the author of the book, *Dams, Migration and Authoritarianism in China: The Local State in Yunnan*, published by Routledge.

Tyler Harlan is an assistant professor in the Department of Urban and Environmental Studies at Loyola Marymount University. He was previously a postdoctoral fellow in sustainability in the Atkinson Center for a Sustainable Future at Cornell University. He received a Ph.D. in Geography from the University of California, Los Angeles, an M.Phil. in Resource Management and Geography from the University of Melbourne, and a B.A. in Anthropology and East Asian Studies from Vanderbilt University. He researches the political economy of China's energy transition, and the implications of this transition for other countries in the region.

Thomas Hennig is a researcher at Philipps-Universität Marburg, Germany. As geographer, for more than 15 years, he puts special attention to various issues related to water management and human-environmental interactions in Asia. Based on three DFG-Projects (German Research Foundation) and in cooperation with Asian International River Centre,

Kunming he gained profound expertise on the consequences and implications of Yunnan's hydropower development.

Darrin Magee is an Associate Professor of Environmental Studies at Hobart and William Smith Colleges in Geneva, New York. He has published numerous articles, book chapters, and other works on water and energy issues in China, and on environmental issues in Asia more broadly.

Franziska Plümmer studies international relations, with a focus on security studies and mobility in China and East Asia. She is a postdoc at the University of Vienna researching Chinese migration and border regulation and looks at how the border becomes a method for Chinese regional development.

Thomas Ptak is an interdisciplinary human-environment geographer whose scholarship critically investigates empirical outcomes of contemporary energy transitions, with a specific focus on policies, practices and planning. His research and teaching critically examine relationships coupling energy and social phenomena with human and non-human environments to pursue salient, pressing questions interrogating access to and exploitation of resources, power—both literally and figuratively—food, water and energy security, development, sustainability and pressing environmental challenges such as human-induced climate change. Tom's scholarship seeks to strengthen a nascent bridge integrating social, physical and geospatial sciences in the burgeoning sub-discipline of energy geography.

Costanza Rampini is an Assistant Professor in the Department of Environmental Studies at San José State University. Her research explores the ways in which dam-building efforts along the Brahmaputra River influence the resilience of downstream communities to climate change impacts on water resources. Her fieldwork takes place in Northeast India in the states of Assam and Arunachal Pradesh. Through her work, Dr. Rampini aims to inform sustainable development and climate action efforts to ensure that they benefit the most vulnerable and underprivileged communities.

Jean-François Rousseau is Assistant Professor at the School of International Development and Global Studies at the University of Ottawa. His research focuses on nature-society relations and addresses the relationships between agrarian change, infrastructure development, and ethnic minority livelihood diversification in Southwest China.

Pon Souvannaseng is Assistant Professor in Comparative & International Policy at Bentley University. She holds a Ph.D. from the London School of Economics. She has conducted fieldwork in Southeast Asia and Africa as a Fulbright scholar and UK Research and Innovation grant funding. Her research interests include international development finance, the political economy of energy infrastructure investment, Third World socialist history and politics, and issues of environmental and economic governance. She has served as a researcher at the UN Research Institute for Social Development, University of Manchester, and as a former Fellow at University College London. She is a Visiting Fellow at the East West Center D.C., a Mansfield Foundation- Henry Luce Asia Scholar and APSA Asia Fellow.

Bryan Tilt is a Professor of Anthropology at Oregon State University. His research focuses on sustainable development, pollution control, and water resources in China and the United States. A former Fulbright Senior Research Scholar in Beijing, he is the author of the book, *Dams and Development in China: The Moral Economy of Water and Power*, published by Columbia University Press.

Andrew van Hulten is a Research Fellow in the School of Humanities and Social Sciences at La Trobe University with a background in Economic Geography. Recent publications can be found in *Impact Assessment and Project Appraisal* and *Economic Geography.*

Brooke Wilmsen has a background in Development Studies with a Ph.D. in Geography. She has worked as a resettlement consultant for several international institutions, government affiliates and private consultancies. She has several years of qualitative and quantitative research experience working on issues of development-forced displacement and resettlement with a focus on the Three Gorges Dam in the People's Republic of China. Brooke has also conducted research on refugee resettlement in Australia.

Abbreviations

ADB	Asian Development Bank
AfDB	African Development Bank
AIIB	Asian Infrastructure Investment Bank
ANEEL	Brazilian Electricity Regulatory Agency
APROSOJA	Mato Grosso State Association of Soy and Corn Producers
BEL	Bujagali Energy Limited Company
BOT	Build-Own-Transfer Arrangement
BRI	Belt and Road Initiative
BRL	Brazilian Real
CCER	The Chinese Certified Emission Reduction
CCP	Chinese Communist Party
CDB	China Development Bank
CDM	Clean Development Mechanism
CHP	Combined Heat and Power Plants
CSG	China State Grid
CSPG	China Southern Power Grid
CTG	The China Three Gorges Corporation
CWE	China International Water & Electric Corporation
CYP	China Yangtze Power
DAF	Development Assistance Fund
DENA	German Energy Agency
DFI	Development Finance Institution
EIA	Environmental Impact Assessment
EIB	European Investment Bank
ELM	Expert Level Mechanism
EPC	Engineering, Procurement, and Construction

EPRDF	Ethiopian People's Revolutionary Democratic Front
GBA	Greater Bay Area
GERD	Grand Ethiopian Renaissance Dam
GHG	Greenhouse Gas
GMS	Greater Mekong Subregion
GoG	Government of Ghana
GoU	Government of Uganda
IBAMA	Brazilian Institute of Environment and Renewable Natural Resources
IDA	International Development Association
IFC	International Finance Corporation
IFI	International Financial Institution
ILO	International Labour Organisation
IMF	International Monetary Fund
IPP	Independent Power Producers
IPS	Industrial Promotion Services Ltd.
IRRM	Impoverishment Risk and Reconstruction Model
LCOE	Levelized Cost of Electricity
LHP	Large Hydropower
LMC	Lancang-Mekong Cooperation
LMRB	Lancang-Mekong River Basin
MIGA	Multilateral Investment Guarantee Agency
MLJY	Meili jiayuan jianshe (*'Constructing a Beautiful Home'*)
MoU	Memorandum of Understanding
MP	Ministério Público (*'Public Prosecutor's Office'*)
MRC	Mekong River Commission
NBA	National Basketball Association
NDRC	National Development and Reform Commission
NFCP	National Forest Conservation Programme
O&M	Operation and Maintenance Costs
OECD	Organisation for Economic Co-operation and Development
OECD-DAC	OECD Development Assistance Committee
PPA	Power Purchase Agreement
PPP	Public-Private Partnership
PRC	People's Republic of China
PRD	Pearl River Delta
PSP	Partnership Support Programme
PV	Photovoltaic
REP	Rural Electrification Programme
RMB	Renminbi
RwD	Resettlement with Development
SGBHL	SG Bujagali Holdings Ltd.
SGC	State Grid Corporation of China

SHP	Small Hydropower
SLA	Sustainable Livelihoods Approach
SOE	State-Owned Enterprise
SPC	State Power Corporation of China's
SPIC	State Power Investment Corp
TGA	The Three Gorges Dam Project Area
UETCL	The Uganda Electricity Transmission Company Limited
USD	US Dollars
W2E	West-to-East
WCD	World Commission on Dams
WTO	World Trade Organisation
YPG	Yunnan Power Grid
YTB	The Yarlung-Tsangpo-Brahmaputra River

List of Figures

List of Tables

CHAPTER 1

Introduction: Southwest China's Hydropower Expansion and Why It Matters There and Beyond

Jean-François Rousseau and Sabrina Habich-Sobiegalla

The People's Republic of China (PRC) sets its own standards when it comes to hydropower. In 2017, the total installed capacity of Chinese hydropower stations reached over 340 GW, up from under 80 GW in 2000 (NBS 2019). The installed capacity of the dams built in China over the last two decades is greater than the combined capacity of all the dams ever built in the United States and Brazil, the world's second- and third-largest hydropower generators (Liu et al. 2018). In 2016, China's

J.-F. Rousseau (✉)
School of International Development and Global Studies,
University of Ottawa, Ottawa, ON, Canada
e-mail: jf.rousseau@uottawa.ca

S. Habich-Sobiegalla
Institute of Chinese Studies, Freie Universität Berlin, Berlin, Germany
e-mail: Sabrina.Habich-Sobiegalla@fu-berlin.de

J.-F. Rousseau and S. Habich-Sobiegalla (eds.), *The Political Economy of Hydropower in Southwest China and Beyond*, International Political Economy Series, https://doi.org/10.1007/978-3-030-59361-2_1

hydropower stations generated close to 1120 TWh of electricity, equivalent to about a fifth of the country's electricity consumption, while 72 per cent of national output came from thermal (mostly coal) plants (NBS 2019). With no short-term replacement for coal in sight, hydropower currently stands out as the only energy source that has the potential to partly contain China's fossil fuel usage, with wind a distant second alternative.

Southwest China, encompassing Sichuan, Guizhou, and Yunnan Provinces together with the Tibet Special Administrative Region and Chongqing Municipality, contains about two-thirds of the nation's 541.6 GW total potential hydropower capacity. Most of this potential emanates from the Himalayas and contiguous mountain ranges along upstream sections of Chinese and transnational rivers. The hydropower potential of this region, known as 'Asia's water tower', long remained underexploited compared to other parts of China (Li et al. 2018), but in the last two decades it has driven an ongoing hydropower development boom. In Yunnan Province alone, hydropower capacity grew from 4 to 60 GW in the twenty years prior to 2016 (Li et al. 2018).

National, provincial, and local government authorities frame hydropower expansion in Southwest China as a green endeavour, taming what are considered to be abundant, underexploited, clean, and renewable energy supplies. Officials cite its benefits for national energy security, rural economic development, and the environment, emphasising reduced greenhouse gas (GHG) emissions and atmospheric pollution (Li and Shapiro 2020). Moreover, local populations in remote southwestern areas that include a high proportion of ethnic minorities are claimed to benefit from electrification and/or economic development, with improved livelihoods (mostly assessed in terms of financial income) and enhanced citizenship 'quality' (*suzhi*).

Aside from such assertions of socio-environmental benefits, official discourses have little to say about the potential impacts of Southwest China's hydropower boom upon ecosystems and populations, the different upstream and downstream consequences of dam construction, and negative long-term impacts (cf. Ritcher et al. 2010). Hydropower is not, in fact, a carbon-neutral energy: reservoirs do emit carbon dioxide and methane, though according to current (incomplete) models, GHG emissions from the hydropower sector are minimal compared to gas and coal thermal plants (Li et al. 2015). Hydropower expansion alters river connectivity, impeding riverine hydrological regimes' capacity to fulfil a host of environmental functions, some of which are vital to socio-economic activities like inland fisheries and floodplain farming (Grill et al. 2019). River fragmentation notably reshuffles water and sediment flows and the seasonal regimes which have customarily sustained riparian farm systems and land capacity in Southwest China and beyond. Dams and reservoirs also create barriers for aquatic life, including along fish migration routes, and alter riverbeds in ways that modify and/or destroy habitats. While river fragmentation levels have long been lower in the Southwest compared to Central and Eastern China, river connectivity here has decreased the fastest in recent decades. Other land-based environmental consequences include increased seismic activity around reservoirs, which some argue triggered the catastrophic 2008 Wenchuan earthquake in Sichuan that led to 70,000 deaths (Chen et al. 2018).

Hydropower expansion also involves the flooding of land in the course of reservoir creation, which triggers land use and land cover changes as well as livelihood reorganisations. By the year 2000, Chinese reservoirs already spanned two million hectares (Vermeer 2011), and their size has grown dramatically since then. According to official figures, 10.2 million people had been resettled in the course of reservoir creation in China between 1949 and 1989; by 2007 that number had climbed to 23 million (Beck et al. 2012; Wilmsen 2016).[1] Over a million people, most of whom lived in Chongqing, Southwest China, were relocated to make room for the mammoth Three Gorges Dam reservoir alone (Wilmsen et al. 2011).[2] Partly in order to better accommodate that resettlement process, Chongqing was bestowed provincial-level municipality status in 1997, illustrating how hydropower expansion yields resource reallocations governed by inherently political processes (Bijker 2007).

The political dimensions of Southwest China's hydropower expansion transcend the country's national boundaries, as many of the dams built here are located along international rivers or their tributaries. Partly due to China's dam-building activities, in August 2019, water levels along the Mekong River dropped to a 100-year low, causing electricity blackouts

[1] When referring to people who were displaced in the course of reservoir creation, the contributors to this volume refer to 'dam migrants', 'resettlers', or 'resettled people' interchangeably.

[2] The Three Gorges Dam itself is located in Yichang in Hubei Province, which is generally considered to be in Central China.

and a 80–90% decline in fish catches in Cambodia, salt water intrusion in Vietnam, and a 1.5 billion US dollar hit to Thailand's GDP (*The Economist* 2019, 2020). This occurs in a context where China holds a disproportionate influence upon regional hydropolitics as well as a geostrategic advantage due to its status as an upstream nation; it is also a geopolitical giant compared to mainland Southeast Asian states, and to a lesser extent India. Relatedly, China's neighbours have long complained that it treats hydropower expansion along international rivers as a solely domestic issue. China has been blamed for refusing to participate in and/or approve of multilateral institutions focused on transboundary water resources such as the Mekong River Commission, the World Commission on Dams, and the 1997 Convention on the Law of Non-Navigational Uses of International Watercourses (Menniken 2007; Nickum 2008).[3] When China has agreed to cooperate with downstream neighbours on issues of river management, it has mostly been in cases where its foreign policy interests would be harmed otherwise (Biba 2014).

Yet criticisms of Southwest China's dams have been less vociferous than one might expect given the magnitude of their potential downstream impacts (Ptak and Hommel 2016). This could be because some of its neighbours are eager to facilitate the capitalist and regionalist projects that drive hydropower expansion in the Southwest and beyond, including Chinese-sponsored high-tension lines, dam development finance, and know-how (Glassman 2010; Su 2013; Ptak and Hommel 2016; Lamb and Dao 2017).

Our Purpose

This book aims to address what we consider a major lack of scholarship on this hydropower spree, which is unique both in its scale and in the importance of the geopolitical, social, and environmental consequences it breeds. To our knowledge, this collection is the first to address this process through various disciplinary and conceptual lenses within the social sciences. We argue that its status as a sensitive (*mingan*) topic explains why Southwest China's hydropower expansion remains under the radar.

[3] China was one of only three nations to vote against the UN convention, besides Turkey and Burundi.

Chinese public intellectuals have long faced hurdles in participating in debates on hydropower expansion, but sensitivities surrounding this topic became more apparent—at least to the outside world—starting in the late 2000s. Although Dai Qing's *Yangtze! Yangtze!* collection on the Three Gorges Dam project was rapidly banned in China after publication in 1989 (see Dai et al. 1994), it nonetheless paved the way for a flourishing of diverse of points of view within the public sphere, mostly in the late 1990s and early 2000s (see Dai 1989; Mertha 2008, 2009). However, this debate abruptly ended when Premier Wen Jiabao decided to halt damming work along the Nu Jiang (Upper Salween) River in 2004.[4] This decree was partly a response to the high-profile opposition of Chinese media, academic, and NGO circles to damming one of Asia's last 'free-flowing rivers'.

In the wake of Premier Wen's decision, many of the China-based activists previously on the front lines of hydropower debates had to move on to other less sensitive endeavours, and could unfortunately not participate in this collection. Chinese scholars face similar disincentives and constraints in researching and documenting the social impacts of hydropower development. We therefore believe it is important to put some of the scholars involved in these debates in conversation with each other. We hope that this collection will encourage further work that better identifies what is at stake along Southwest China's rivers—and explains why it matters to academic and policy audiences as well as the broader public.

We review some of this scholarship in the remainder of the introduction, first engaging with literature that allows us to conceptualise the ongoing hydropower expansion in Southwest China as a sociopolitical project transcending the construction of dams. We then outline the contours of the political economy that shapes Southwest China's hydropower boom, and contextualise the region and the social outlook that characterises it. The last section introduces the individual chapters that comprise this collection.

[4] The fate of the Nu River dams remains unclear. While five projects were enshrined in the Twelfth Five-Year Plan (2011–2015), suggesting that they would proceed (Tilt 2014), media reports in 2016 confirmed that the dams had been shelved again (Phillips 2016).

Dams as Projections of Modernity?

Hydrological schemes such as irrigation projects, dikes, canals, and hydropower dams have long been posited as vectors for specific sociopolitical projects. This is chiefly because controlling water, a substance that gathers in cavities and flows downward along continuous beds, requires massive labour and/or financial investments. Wittfogel (1957) asserted that centralised management is a prerequisite to large-scale hydraulic infrastructure development, and that the massive hydraulic works he saw in Asia (and beyond) testified to the despotic inclinations of the rulers who commissioned these works. Yet critics have long rejected this approach as deterministic, and numerous factual counterexamples have been introduced to challenge it (Banister 2014). Wittfogel's thesis nonetheless set the scene for social scientists to probe the complex ways in which water control illustrates specific power relations, and to overcome the nature-society dichotomy that was present in early hydropolitical works (Banister 2014; Linton and Budds 2014; Rogers et al. 2016).

Dams in particular have been closely associated with state actors' modern—or 'high-modernist' (Scott 1998)—projects to make social and environmental matters legible, constant, and predictable, and thus easier to control (Bakker 2012). This legibility is core to state-making processes and is an expected outcome of the territorial reorganisations and resource reallocations that accompany hydropower expansion (Scott 1998; Tilt 2014). Hydropower expansion can be understood as a territorial and extractive endeavour that utilises specific technological means and methodologies to install a particular governance regime over land and water resources as well as riparian populations (cf. Vandergeest 1996; Rogers et al. 2016; Rogers and Wilmsen 2020). This territorialisation is meant to consolidate specific social and spatial hierarchies, together with the world views and ideologies supporting them (Swyngedouw 2015; Rogers and Crow-Miller 2017). This illustrates how the penetration of new technologies and the arguments that support it are inherently political activities, though the allegedly 'objective' policy and technical documents that accompany hydropower expansion attempt to depoliticise dams and their impacts (cf. Bijker and Law 1992; Latour 2002).

Waterscapes are the sites of the material and metaphorical connections that hydropower development triggers and the loci of the discursive practices supporting such development (Swyngedouw 1999; Baviskar 2007; Orlove and Caton 2010). Waterscapes thus encompass the complexities of interfaces among water, land, farmers, fisherfolk, urban hydroelectricity users, cadres from different administrative levels, corporate stakeholders, NGO representatives, and so on. In such settings, changes in the qualitative and quantitative attributes of water may create access opportunities or trigger exclusion processes that themselves drive cascading benefits and consequences across spatial and temporal scales (Mehta et al. 2012; Franco et al. 2013). These variables manifest differently for specific actors and can simultaneously alter or enhance water's capacity to sustain diverse and complex socio-ecological functions (Bakker 2010, 2012).

Typically, powerful government and corporate actors advocating for hydropower expansion reap the greatest benefits from this activity, whereas rural riparian populations whose livelihoods are often centred on customary hydrological regimes experience the worst impacts. The actors that advocate for hydropower expansion therefore deploy a wide range of practices and rationalities to legitimise such outcomes, drawing on discursive strategies, scientific evidence, and legal and policy instruments (Hall et al. 2011; Rogers and Crow-Miller 2017). Taken together, these technologies of government pave the way for governmentality regimes to shape 'the conduct of conduct' in a given waterscape (Foucault 2002: 341). In the process, new subjects are actively created to align with the state programme that fosters hydropower expansion in the first place (Rogers and Wilmsen 2020; Habich-Sobiegalla and Rousseau 2020). Yet ethnographic case studies demonstrate that local populations often use their agency to develop overt and/or covert strategies that do not fully align with the social project planned for them (Habich 2016).

The Political Economy of Southwest China's Hydropower Expansion

Hydropolitics in Southwest China

China has a long history of water control and hydraulic infrastructure development, as the capacity to avoid natural catastrophes, including floods and drought, determined whether dynasties maintained or lost the 'mandate of heaven' during the imperial period. Among this infrastructure was the massive Grand Canal, completed in the fourteenth century, which secured grain transport between the kingdom's rice baskets in the South and the northern capital. Centuries later, its contemporary equivalent, the South-to-North Water Diversion project, continues this emphasis on hydraulic technologies as developmental solutions (Tilt 2014; Pietz 2015). In this specific case, rerouting portions of the Yangtze River's discharge is meant to alleviate a lack of water resources in the North. Underscoring this and other mammoth hydraulic infrastructure projects in China is the anthropocentric belief of all generations of modern Chinese leaders that nature, including water resources, should serve the regime's ideology—whether Maoism or 'socialism with Chinese characteristics' (cf. Shapiro 2001).

Dams have always been central to the PRC's state-building and socio-economic agendas, and research on hydroelectricity has consequently yielded important insights into governance processes that transcend this specific sector. Most notably, building on case studies including the Three Gorges Dam, Lieberthal and Oksenberg (1988) coined the expression 'fragmented authoritarianism' in reference to the bargaining processes spanning both horizontal and vertical lines of control within the state apparatus, which see state actors of various administrative levels vie to implement their sometimes conflicting agendas. The Three Gorges Dam highlighted these tensions perhaps better than any other project in China. First envisioned by Sun Yat-sen in the 1930s and the subject of poetry by Mao a few decades later, the project went through multiple cycles of shelving and resurfacing before the National People's Congress approved it in 1992. Yet even then the project was the Chinese parliament's most contested vote ever (Dai et al. 1994).

Mertha (2008, 2009) has expanded the fragmented authoritarianism framework to account for new non-state 'policy entrepreneurs' that have joined the hydropower governance regime, including NGO campaigners and public intellectuals. While we argue that civil society actors' participation in governance was a short-lived phenomenon, the pluralisation of Chinese hydropolitics is also seen in the rising importance of corporate actors and logics. The State Power Corporation of China's monopoly over electricity generation and transmission was dismantled in 2004, with electricity generation portfolios allocated to five main public conglomerates (Yeh and Lewis 2004; Andrews-Speed 2012).[5] These 'big five' producers and a host of smaller 'Independent Power Producers', either linked to public utility firms or not, have long been eager to expand

[5] These groups are China Huaneng Group, China Datang Corporation, China Huadian Corporation, China Guodian Corporation, and China Power Investment Corporation (now State Power Investment Corporation).

their capacity and their respective market shares. Electricity generation, particularly hydroelectricity, may now face an oversupply (Hennig and Harlan 2018), even with the rapid proliferation of high-tension lines connecting Southwest China to energy-hungry markets along China's eastern coast and to export markets such as Laos and Vietnam (Hennig et al. 2016). In any case, these corporate actors have now become full-fledged—and very powerful—participants in the governance processes that oversee hydropower expansion in Southwest China.

The 'powershed', a notion coined by Darrin Magee (2006), encompasses the social, political, economic, and environmental contexts of hydropower development and electricity delivery. This concept is useful in analyses of the diversity of Chinese social actors and territory, allowing for the dissection of the discourses and policies that allocate very specific socio-environmental functions to the East and the Southwest. Within the powershed, the East is the wealthy and developed locus of political and economic power, the decision centre for hydropower developments in the Southwest, and the main market for the Southwest's hydroelectricity. In contrast, the Southwest is framed as resource-rich but underdeveloped, with local populations, livelihoods, and ethnic minorities portrayed as 'backward' (Magee 2006). Hydropower development and its social, economic, and environmental impacts become normalised, and even desirable, in this context (cf. Yeh 2013).

The 'Open up the West' campaign (*xibu da kaifa*) was a key policy mechanism introduced in 1999 to address growing developmental inequalities between eastern and southwestern regions by channelling investment between them (Goodman 2004). The programme notably aimed to tap into the remote West and its resources, including hydropower, to fuel growth both in the West and in the East. The campaign, together with related policy measures such as 'Send Western Electricity East' (*xidian dongsong*), thus centred chiefly on infrastructure development. Regional and local authorities were sent a strong signal that they would be judged by their achievements in attracting hydropower investment (Magee 2006; Glassman 2010).

Besides regional inequalities, atmospheric pollution is another externality of development that has steadily gained prominence in the Chinese regime's policy priorities. The Renewable Energy Law enacted in 2005 and amended in 2009 sets ambitious hydroelectricity capacity development objectives (Schuman and Lin 2012). Fifteen years on, the mandatory targets of the law have not just been attained but indeed surpassed, and the renewable energy sector has now emerged as China's flagship 'low-carbon industry' (Harlan 2018). Actively promoted by the state, global carbon finance instruments like the Clean Development Mechanism (CDM) enshrined in the Kyoto Protocol, as well as national carbon markets like the Chinese Certified Emission Reduction (CCER), have been embraced by hundreds of Southwest China-based dam projects (Bayer et al. 2013; Lo and Cong 2017). It remains to be seen, however, whether carbon financing has truly driven additional capacity expansion in the Southwest compared to what would have prevailed under 'business as usual' scenarios.

The above policy priorities all emanate from the central government, which sets broad orientations and goals but provides little guidance on how these should be attained and monitored. Implementation is left to lower-level authorities, who often lack resources and expertise (Habich 2016). Case studies of communities resettled due to hydropower expansion highlight how top-down resettlement guidelines seldom achieve their objective to enhance migrant livelihoods, instead often leading to livelihood degradation (see Habich 2016; Wilmsen 2016). The unequal power relations bred by hydropower expansion underpin challenges in providing rural resettlers with proper landholdings (Rogers and Wilmsen 2020). Local cadres are incentivised to ensure that hydropower expansion unfolds as quickly and cheaply as possible, with proper application of resettlement guidelines taking a back seat (Habich 2015).

Places, People, and Red Stamps

Southwestern China covers about a quarter of the country's landmass and hosts a population of 200 million, slightly less than 15% of the national total (NBS 2019). Half of the Southwest's residents are registered as rural dwellers, compared to 41.5% nationally (NBS 2019); the average per capita disposable income is about 25% lower than the countrywide average and less than half that of eastern areas. The Southwest is also China's most ethnically diverse region. Two-thirds of the nation's 55 official ethnic minorities have historically settled in the area, and ethnic minorities comprise 20% of the regional population compared to 8.5% nationally.

Much of Southwest China belongs to the transnational region known as the Southeast Asian Massif and/or Zomia (Scott 2009; Michaud et al. 2016). Diverse ethnic minorities have historically taken advantage of the

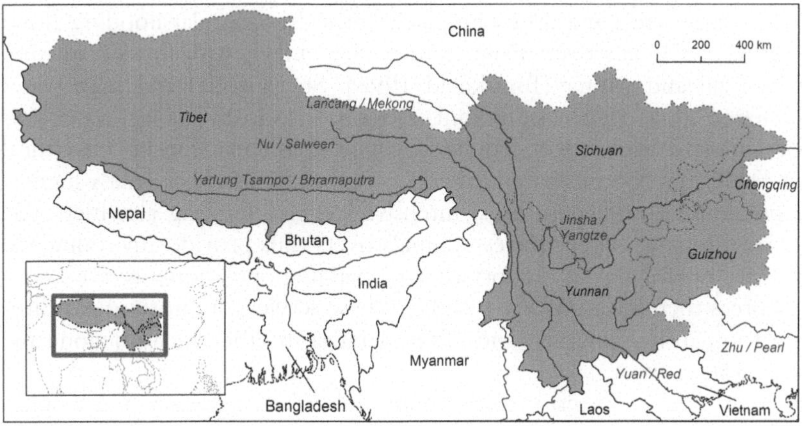

Fig. 1.1 Southwest China and its main rivers

'friction of terrain' in this mountainous, peripheral area, posing challenges for the state-making endeavours of Chinese emperors as well as French and English colonial governments (Giersch 2006; Scott 2009). Nowadays, one of China's strategies to overcome this friction and develop the frontier is to encourage projects involving the cooperation of adjacent countries (Su 2013). Energy infrastructures such as transnational high-tension lines and pipelines are central to this endeavour (Ptak and Hommel 2016; Su 2016). Since 2000, the central government's 'Going Out' policy has encouraged state-owned enterprises to invest overseas, and the more recent 'One Belt One Road' scheme—deemed China's Marshal Plan—further fosters Chinese investment in Southeast Asia-based hydropower schemes, notably in Myanmar, Laos, and Vietnam (Andrews-Speed et al. 2016; Yeh 2016; Lamb and Dao 2017).

Southwest China hosts the headwaters of six major river systems characterised by steep upstream watersheds, wide lowland delta floodplains, and significant seasonal hydrological variation (Dudgeon 1992). The case studies in this collection address hydroelectric developments implemented along the Yarlung Tsangpo/Brahmaputra (YTB), Nu/Salween, Lancang/Mekong, Red, and Jinsha/Yangtze watersheds (Fig. 1.1).[6]

[6] The Pearl River comprises the sixth major watershed.

Monsoon rains and upstream glacier melt drive regular flooding along Southwest China's rivers from April to September, with flows retreating in the fall and winter (Brookfield 1998). Steep slopes and high water discharge also trigger significant erosion, seasonally filling waterways with massive amounts of sediment. These sediment deposits are critical for fertilising floodplains and deltas, some of which host Asia's highest population densities and most intensive wet-rice farming activities. The seasonal regimes of Southwest China's rivers thus provide environmental services potentially as important to riparian populations as the water resources themselves (Wolters 2007). These services are notoriously difficult to quantify, however, and often manifest in different ways along the same waterway.

Hydropower expansion significantly reshuffles customary hydrological regimes when reservoirs annihilate seasonal water discharge variations and trap sediment. The sediment trapped in reservoirs is a concern for downstream farmers and fisherfolk and for dam operators alike, since reservoir siltation impedes efficiency (cf. Yu et al. 2019). Partly in order to address this issue, hydropower development in Southwest China typically follows a so-called cascade model, where smaller upstream reservoirs serve to trap sediment and increase the efficiency of more powerful downstream hydroelectric stations (Hennig et al. 2013).

Major dam projects must obtain 'red stamps' from various administrative bureaus, including the Ministry of Water Resources and the National Development and Reform Commission, prior to construction (Habich 2016). Smaller dams necessitate lower-level approvals than larger ones; small hydropower schemes ($\leq$50 MW installed capacity[7]) typically only require county-level approval, medium ones (50–300 MW) must obtain provincial-level clearance, and large projects ($\geq$ 300 MW) are validated at the national level (Magee 2006; Hensengerth 2010, 2014). In addition, factors specific to individual dam projects can influence such administrative procedures. For instance, dams planned for international watersheds are subject to higher-level scrutiny than their strictly domestic equivalents.

Since the implementation of the Environmental Impact Assessment (EIA) law in 2003, EIAs have been required for development projects

[7] These values represent the Chinese categorisation of hydroelectric stations. Globally, 10 MW is the more often recognised threshold to distinguish small from medium-sized hydropower stations; dams higher than 15 m are usually considered large dams (cf. Hennig and Harlan 2018).

expected to trigger significant social or environmental impacts, which often exclude small hydropower stations (McDonald 2007; Tilt 2014). Accredited consulting firms are responsible for drafting the assessments; together with the fact that only EIA summaries must be publicly released, this has been observed to create bias in favour of dam projects (Tilt 2014). Though the EIA law emphasises public stakeholder consultations, no specific mechanism exists to ensure that participants are granted enough information about proposed projects to comment on them, nor that a diversity of opinions is accounted for during assessment processes (Yang 2008). In spite of these caveats, the EIA law may have influenced the decision to halt flagship dam projects along the Nu and Jinsha (Upper Yangtze) Rivers (Hennig et al. 2013)—though perhaps only temporarily.

Resettlement schemes in China are not governed by any particular law, but instead by sector-specific regulations (*tiaoli*) on resettlement planning, implementation, and compensation related to water resource development projects. These regulations were published first in 1991 and then updated in 2006 and 2017; a series of related regulations, opinions (*yijian*), and measures (*banfa*) have followed, introducing ever more detailed guidelines on how to effectively manage resettlement schemes and growing numbers of dam migrants. These guidelines are implemented at lower levels of government, where central stipulations have their details filled in and adapted to local circumstances (*yindi zhiyi*). Based on this framework, specific resettlement plans are designed by provincial-level government departments, design agencies, and hydropower companies for each large dam built within the province (see Hensengerth 2010; Habich 2016).

The political-economic circumstances described here all foster hydroelectric development in Southwest China. Yet these rationales do not necessarily align with those of local riparian populations, for whom land is a core asset and agrarian activities remain an important livelihood component. Though local populations—including ethnic minorities—might welcome infrastructure or agrarian expansion projects and their concomitant economic opportunities, hydropower dams are seen less positively (cf. Litzinger 2007). Hydropower expansion triggers simultaneous land and water resource reallocations that often result in local populations' permanent exclusion from their land- and water-based livelihood activities (Rousseau 2019). As this occurs, work migration becomes an increasingly common coping mechanism for local populations (Galipeau et al. 2013; Tilt 2014). While migration allows dam-affected households to access greater financial resources, the accompanying social changes can decrease social capital levels within village communities (Tilt and Gerkey 2016). For those who maintain some access to land, agricultural intensification and marketisation are frequent strategies; these adaptations can also drive undesirable social changes, including agricultural specialisation and the rapid expansion of chemical fertiliser-dependent farm systems (Rousseau 2017).

Besides these livelihood impacts, earlier scholarship has documented how local authorities often fail to implement compensation guidelines thoroughly, resulting in dam-affected households obtaining less financial or material compensation than what policies stipulate. For instance, the value of crops grown within soon-to-be-flooded areas is sometimes underestimated to depress compensation payments (Rousseau 2017).[8] The substandard land and housing that resettled communities receive can likewise trigger resentment, with complaints that local authorities fail to ensure households' legal rights are being protected (Habich 2016; Rousseau et al. 2017). Widespread corruption incentivises local cadres to keep resettlement costs to a minimum for hydropower companies and ensure that dam development proceeds swiftly.

Hydropower expansion also fosters broader non-material sociocultural changes within affected communities. As flooded land disappears, so too do the multi-generational relationships of its users with their specific landholdings. Unlike many societies in the adjacent midlands and highlands who have customarily been involved in swidden farming, riparian communities in lowland river valleys have often been traditionally sedentary, relying on the flood and pulse regime to fertilise their lands regularly. Specific plots of land have therefore emerged as a core component of their identity; the disappearance of these landholdings is a deeply traumatic event (Michaud 2011). Hydroelectric expansion relatedly often spearheads a series of technological changes within dam-affected communities, including easier access to transport infrastructure, satellite dishes, and modern farm equipment; all of these technologies foster the 'modernisation' of rural cultures and the adoption of lifestyles increasingly aligned with the state's social agenda (cf. Turner et al. 2015; Ptak, this volume).

[8] All land remains state property in China, and rural households own land usage rights. The value of the crops grown in a given area is thus a proxy for fixing compensation payments, with villagers paid for lost income opportunities rather than for the land itself.

The Sections and Chapters in This Collection

This collection is organised around three sections that cover several aspects of hydropower development in Southwest China: hydropower and resettlement governance; dams and rural livelihoods; and transnational matters. The first section on hydropower and resettlement governance commences with a chapter by Thomas Hennig and Darrin Magee highlighting the main trends that have shaped hydropower production and consumption in Southwest China since the early 2000s, together with the role that grid infrastructure deployment has played in these evolutions. The next two chapters document how and why the construction of the Three Gorges Dam compelled the central government to make its resettlement policies increasingly people-oriented, together with local policy innovations that emerged at the onset of central-level policy realignments. Brooke Wilmsen, Yuefang Duan, and Andrew van Hulten explore the pathways and motivations of resettlers who left Three Gorges Dam resettlement sites. Bettina Gransow investigates the challenges and long-term coping strategies of a village community that underwent distant resettlement from the Three Gorges area to the Pearl River Delta; she highlights differences in perceptions and normative frameworks regarding resettlement among the displaced people, the local host village, and the local host government. The next chapter by Bryan Tilt and Zhuo Chen is about the effects of dam-induced displacement along the Lancang River. Their findings suggest that while government and industry efforts to minimise harm to displaced communities are improving, more consideration of social and cultural impacts, property rights, and access to natural resources is needed. The final chapter of this section is by Sabrina Habich-Sobiegalla and Franziska Plümmer, who discuss the *meili jiayuan jianshe* village renovation scheme, a recently introduced component of resettlement policies. Their contribution demonstrates how the state actively mobilises infrastructure development as a strategy to reintegrate dam migrants into national developmental discourses, and shows that this occurs at the cost of further marginalisation for displaced populations.

The second section on dams and rural livelihoods focuses on small- and medium-sized dams in particular. The first chapter by Tyler Harlan is on the social benefits of small hydropower in China; he finds that it is neither green nor directly beneficial to poor people, but that its impacts hinge on broader issues of energy affordability and power sector profitability. The following chapter by Thomas Ptak provides insights into how the development of small hydroelectric operations and electrification have operated as a Trojan horse, leading to the wide diffusion of electric consumer goods that have fostered the 'modernisation' of small, rural ethnic minority communities. Jean-François Rousseau then goes on to investigate how hydropower expansion and its long-term influences upon livelihood asset portfolios shape the strategies that riparian populations have been pursuing after the construction of a dam.

Finally, the third section on transnational and international matters commences with a chapter by Darrin Magee, who argues that hydropower could positively balance intermittent, non-dispatchable renewable energy technologies like wind and solar in China and beyond, though achieving this potential would require governance reforms that have yet to manifest. The chapter by Sebastian Biba applies the notion of 'hydro-hegemony' in the context of the Mekong River, investigating power asymmetries between China and other riparian countries as well as instances of cooperation and domination along Southeast Asia's most important waterway. Costanza Rampini then looks at the YTB system, exploring how China and India jointly govern this transboundary watershed and shedding light on events shaping Sino-Indian tensions. She highlights how China and India do cooperate over water along the YTB, and finds that climate change impacts on the river and the glaciers that feed it are among the many tests that this collaboration is now facing. The chapter by Pon Souvannaseng discusses contemporary patterns of Chinese-supported hydropower development across Africa, looking in depth at the ideational, fiscal, and historical factors driving the turn to Sino-finance in dam development in Ghana and Uganda. Finally, the chapter by Ricardo Andrade, set in the Brazilian Amazon, uncovers how Chinese interests shape development processes where hydropower development, fluvial transport, and soybean trade intersect. Andrade demonstrates how an obscure legal instrument dating from the military dictatorship facilitates these developments and ensures that Brazil does not miss out on the global Chinese investment bonanza.

References

Andrews-Speed, P. (2012). *The governance of energy in China: Transition to a low-carbon economy*. Basingstoke: Palgrave Macmillan.

Andrews-Speed, P., Qiu, M., & Len, C. (2016). Chinese engagement in Southeast Asian energy and mineral resources: Motivations and outlook. *Eurasian Geography and Economics, 57*(3), 316–342.

Bakker, K. (2010). *Privatizing water: Governance failure and the world's urban water crisis*. Ithaca: Cornell University Press.

Bakker, K. (2012). Water: Political, biopolitical, material. *Social Studies of Science, 42*(4), 616–623.

Banister, J. M. (2014). Are you Wittfogel or against him? Geophilosophy, hydrosociality, and the state. *Geoforum, 57*, 205–214.

Baviskar, A. (Ed.). (2007). *Waterscapes: The cultural politics of a natural resource*. New Delhi: Permanent Black.

Bayer, P., Urpelainen, J., & Wallace, J. (2013). Who uses the clean development mechanism? An empirical analysis of projects in Chinese provinces. *Global Environmental Change, 23*(2), 512–521.

Beck, M. W., Claassen, A. H., & Hundt, P. J. (2012). Environmental and livelihood impacts of dams: Common lessons across development gradients that challenge sustainability. *International Journal of River Basin Management, 10*(1), 73–92.

Biba, S. (2014). Desecuritization in China's behavior towards its transboundary rivers: The Mekong river, the Brahmaputra river, and the Irtysh and Ili rivers. *Journal of Contemporary China, 23*(85), 21–43.

Bijker, W. E. (2007). Dikes and dams, thick with politics. *Isis, 98*(1), 109–123.

Bijker, W. E., & Law, J. (1992). *Shaping technology/building society: Studies in sociotechnical change*. Cambridge: MIT Press.

Brookfield, M. E. (1998). The evolution of the great river systems of southern Asia during the Cenozoic India-Asia collision: Rivers draining southwards. *Geomorphology, 22*, 285–312.

Chen, Y., Hu, J., & Peng, F. (2018). Seismological challenges in earthquake hazard reductions: Reflections on the 2008 Wenchuan earthquake. *Science Bulletin, 63*(17), 1159–1166.

Dai, Q. (Ed.). (1989). *Yangtze Yangtze—Controversy over the Three Gorges Dam project*. Guiyang: Guizhou People's Publishing House.

Dai, Q., Adams, P., & Thibodeau, J. (Eds.). (1994). *Yangtze! Yangtze!*. London: Earthscan.

Dudgeon, D. (1992). Endangered ecosystems: A review of the conservation status of tropical Asian rivers. *Hydrobiologia, 248*(3), 167–191.

Foucault, M. (2002). *Power (The essential works of Foucault, 1954–1984, vol. 3)* (Ed.: Faubion, J. D.). London: Penguin.

Franco, J., Mehta, L., & Veldwisch, G. J. (2013). The global politics of water grabbing. *Third World Quarterly, 34*(9), 1651–1675.

Galipeau, B. A., Ingman, M., & Tilt, B. (2013). Dam-induced displacement and agricultural livelihoods in China's Mekong Basin. *Human Ecology, 41*(3), 437–446.

Giersch, C. P. (2006). *Asian borderlands: The transformation of Qing China's Yunnan frontier*. Cambridge: Harvard University Press.

Glassman, J. (2010). *Bounding the Mekong: The Asian development bank, China, and Thailand*. Honolulu: University of Hawai'i Press.

Goodman, D. S. G. (2004). The campaign to "Open up the west": National, provincial-level and local perspectives. *The China Quarterly, 178,* 317–334.

Grill, G., Lehner, B., Thieme, M., Geenen, B., Tickner, D., Antonelli, F. et al. (2019). Mapping the world's free-flowing rivers. *Nature, 569*(7755), 215–221.

Habich, S. (2015). Strategies of soft coercion in Chinese dam resettlement. *Issues & Studies, 51*(1), 165–199.

Habich, S. (2016). *Dams, migration and authoritarianism in China: The local state in Yunnan*. New York: Routledge.

Habich-Sobiegalla, S., & Rousseau, J.-F. (2020). Responsibility to choose: Governmentality in China's participatory dam resettlement processes. *World Development, 135,* 1–11.

Hall, D., Hirsch, P., & Murray Li, T. (2011). *Powers of exclusion: Land dilemmas in Southeast Asia*. Singapore: NUS Press.

Harlan, T. (2018). Rural utility to low-carbon industry: Small hydropower and the industrialization of renewable energy in China. *Geoforum, 95,* 59–69.

Hennig, T., Wang, W., Magee, D., & He, D. (2016). Yunnan's fast-paced large hydropower development: A powershed-based approach to critically assessing generation and consumption paradigms. *Water, 8*(10), 476.

Hennig, T., & Harlan, T. (2018). Shades of green energy: Geographies of small hydropower in Yunnan, China and the challenges of over-development. *Global Environmental Change, 49,* 116–128.

Hennig, T., Wang, W., Feng, Y., Ou, X., & He, D. (2013). Review of Yunnan's hydropower development. Comparing small and large hydropower projects regarding their environmental implications and socio-economic consequences. *Renewable and Sustainable Energy Reviews, 27,* 585–595.

Hensengerth, O. (2010). *Sustainable dam development in China between global norms and local practices*. Bonn: Deutsches Institut für Entwicklungspolitik.

Hensengerth, O. (2014). Between local and global norms: Hydropower policy reform in China. In W. Scheumann & O. Hensengerth (Eds.), *Evolution of dam policies: Evidence from the big hydropower states* (pp. 55–93). Heidelberg: Springer.

Lamb, V., & Dao, N. (2017). Perceptions and practices of investment: China's hydropower investments in Vietnam and Myanmar. *Canadian Journal of*

Development Studies/Revue Candienne D'études Du Developpement, 38(3), 395–413.

Latour, B. (2002). Morality and technology: The end of the means. *Theory, Culture & Society, 19*(5–6), 247–260.

Li, S., Zhang, Q., Bush, R. T., & Sullivan, L. A. (2015). Methane and CO_2 emissions from China's hydroelectric reservoirs: A new quantitative synthesis. *Environmental Science and Pollution Research, 22*(7), 5325–5339.

Li, X., Chen, Z., Fan, X., & Cheng, Z. (2018). Hydropower development situation and prospects in China. *Renewable and Sustainable Energy Reviews, 82,* 232–239.

Li, Y., & Shapiro, J. (2020). *China Goes Green: Coercive Environmentalism for a Troubled Planet*. Cambridge: Polity.

Lieberthal, K., & Oksenberg, M. (1988). *Policy making in China: Leaders, structures, and processes*. Princeton: Princeton University Press.

Linton, J., & Budds, J. (2014). The hydrosocial cycle: Defining and mobilizing a relational-dialectical approach to water. *Geoforum, 57,* 170–180.

Litzinger, R. (2007). In search of the grassroots: Hydroelectric politics in Northwest Yunnan. In E. J. Perry & M. Goldman (Eds.), *Grassroots political reform in contemporary China* (pp. 282–299). Cambridge: Harvard University Press.

Liu, B., Liao, S., Cheng, C., Chen, F., & Li, W. (2018). Hydropower curtailment in Yunnan Province, southwestern China: Constraint analysis and suggestions. *Renewable Energy, 121,* 700–711.

Lo, A. Y., & Cong, R. (2017). After CDM: Domestic carbon offsetting in China. *Journal of Cleaner Production, 141,* 1391–1399.

Magee, D. (2006). Powershed politics: Yunnan hydropower under Great Western Development. *The China Quarterly, 185,* 23–41.

McDonald, K. N. (2007). *Damming China's grand Canyon: Pluralization without democratization in the Nu River Valley*. PhD dissertation, University of California, Berkeley.

Menniken, T. (2007). China's performance in international resource politics: Lessons from the Mekong. *Contemporary Southeast Asia, 29*(1), 97–120.

Mertha, A. (2008). *China's water warriors: Citizen action and policy change*. Ithaca: Cornell University Press.

Mertha, A. (2009). "Fragmented authoritarianism 2.0": Political pluralization in the Chinese policy process. *The China Quarterly, 200,* 995–1012.

Mehta, L., Jan Veldwisch, G., & Franco, J. (2012). Introduction to the special issue: Water grabbing? Focus on the (re)appropriation of finite water resources. *Water Alternatives, 5*(2), 193–207.

Michaud, J. (2011). Conclusion: Lesson for the future. In J. Michaud & T. Forsyth (Eds.), *Moving mountains: Ethnicity and livelihoods in highland China, Vietnam, and Laos* (pp. 215–237). Vancouver: UBC Press.

Michaud, J., Swain, M. B., & Barkataki-Ruscheweyh, M. (2016). *Historical dictionary of the peoples of the Southeast Asian Massif*. Lanham: Rowman & Littlefield.

Nickum, J. (2008). The upstream superpower: China's international rivers. In O. Varis, C. Tortajada, & A. Biswas (Eds.), *Management of transboundary rivers and lakes* (pp. 227–244). Berlin: Springer-Verlag.

National Bureau of Statistics (NBS). (2019). *China statistical yearbook 2018*. http://www.stats.gov.cn/tjsj/ndsj/2018/indexeh.htm. Accessed 2 July 2020.

Orlove, B. S., & Caton, S. C. (2010). Water sustainability: Anthropological approaches and prospects. *Annual Review of Anthropology, 39*, 401–415.

Phillips, T. (2016, December 2). Joy as China shelves plans to dam 'angry river' *The Guardian*, https://www.theguardian.com/world/2016/dec/02/joy-as-china-shelves-plans-to-dam-angry-river. Accessed 2 July 2020.

Pietz, D. A. (2015). *The Yellow River: The problem of water in modern China*. Cambridge: Harvard University Press.

Ptak, T., & Hommel, D. (2016). The trans-political nature of Southwest China's energy conduit, Yunnan Province. *Geopolitics, 21*(3), 556–578.

Ritcher, B. D., Postel, S., Revenga, C., Scudder, T., Lehner, B., Churchill, A., et al. (2010). Lost in development's shadow: The downstream human consequences of dams. *Water Alternatives, 3*(2), 14–42.

Rogers, S., Barnett, J., Webber, M., Finlayson, B., & Wang, M. (2016). Governmentality and the conduct of water: China's South-North Water transfer project. *Transactions of the Institute of British Geographers, 41*(4), 429–441.

Rogers, S., & Crow-Miller, B. (2017). The politics of water: A review of hydropolitical frameworks and their application in China. *WIREs Water, 4*(6), e1239.

Rogers, S., & Wilmsen, B. (2020). Towards a critical geography of resettlement. *Progress in Human Geography, 44*(2), 256–275.

Rousseau, J.-F. (2017). Does carbon finance make a sustainable difference? Hydropower expansion and livelihood trade-offs in the Red River valley, Yunnan Province, China. *Singapore Journal of Tropical Geography, 38*(1), 90–107.

Rousseau, J.-F. (2019). When land, water and green-grabbing cumulate: Hydropower expansion, livelihood resource reallocation and legitimisation in southwest China. *Asia Pacific Viewpoint, 61*(1), 134–146.

Rousseau, J.-F., Orange, D., Habich-Sobiegalla, S., & Nguyen, V. T. (2017). Socialist hydropower governances compared: Dams and resettlement as experienced by Dai and Thai societies from the Sino-Vietnamese borderlands. *Regional Environmental Change, 17*(8), 2409–2419.

Schuman, S., & Lin, A. (2012). China's renewable energy law and its impact on renewable power in China: Progress, challenges and recommendations for improving implementation. *Energy Policy, 51,* 89–109.
Scott, J. C. (1998). *Seeing like a state: How certain schemes to improve the human condition have failed*. New Haven: Yale University Press.
Scott, J. C. (2009). *The art of not being governed: An anarchist history of upland Southeast Asia*. New Haven: Yale University Press.
Shapiro, J. (2001). *Mao's war against nature: Politics and the environment in Revolutionary China*. Cambridge: Cambridge University Press.
Su, X. (2013). From frontier to bridgehead: Cross-border regions and the experience of Yunnan, China. *International Journal of Urban and Regional Research, 37*(4), 1213–1232.
Su, X. (2016). Repositioning Yunnan: Security and China's geoeconomic engagement with Myanmar. *Area Development and Policy, 1*(2), 178–194.
Swyngedouw, E. (1999). Modernity and hybridity: Nature, regeneracionismo, and the production of the Spanish waterscape, 1890–1930. *Annals of the Association of American Geographers, 89*(3), 443–465.
Swyngedouw, E. (2015). *Liquid power: Contested Hydro-modernities in twentieth-century Spain*. Cambridge: MIT Press.
The Economist. (2019, August 7). Why are water levels of the Mekong at a 100-year low? *The Economist*, https://www.economist.com/graphic-detail/2019/08/07/why-are-water-levels-of-the-mekong-at-a-100-year-low. Accessed 2 July 2020.
The Economist. (2020, May 14). South-East Asia's biggest river is drying up. *The Economist*, https://www.economist.com/asia/2020/05/14/south-east-asias-biggest-river-is-drying-up. Accessed 2 July 2020.
Tilt, B. (2014). *Dams and development in China: The moral economy of water and power*. New York: Columbia University Press.
Tilt, B., & Gerkey, D. (2016). Dams and population displacement on China's Upper Mekong River: Implications for social capital and social-ecological resilience. *Global Environmental Change, 36,* 153–162.
Turner, S., Bonnin, C., & Michaud, J. (2015). *Frontier livelihoods: Hmong in the Sino-Vietnamese borderlands*. Seattle: University of Washington Press.
Vandergeest, P. (1996). Mapping nature: Territorialization of forest rights in Thailand. *Society & Natural Resources, 9*(2), 159–175.
Vermeer, E. B. (2011). The benefits and costs of China's hydropower: Development or slowdown? *China Information, 25*(1), 3–32.
Wilmsen, B. (2016). After the deluge: A longitudinal study of resettlement at the Three Gorges Dam, China. *World Development, 84,* 41–54.
Wilmsen, B., Webber, M., & Duan, Y. (2011). Involuntary rural resettlement: Resources, strategies, and outcomes at the Three Gorges Dam, China. *The Journal of Environment & Development, 20*(4), 355–380.

Wittfogel, K. A. (1957). *Oriental despotism: A comparative study of total power*. New Haven: Yale University Press.

Wolters, W. (2007). Geographical explanations for the distribution of irrigation institutions: Cases from Southeast Asia. In P. Boomgaard (Ed.), *A world of water: Rain, rivers and seas in Southeast Asian histories* (pp. 209–234). Singapore: NUS Press.

Yang, S. (2008). Public participation in the Chinese environmental impact assessment (EIA) system. *Journal of Environmental Assessment Policy and Management, 10*(1), 91–113.

Yeh, E. T. (2013). *Taming Tibet: Landscape transformation and the gift of Chinese development*. Ithaca: Cornell University Press.

Yeh, E. T. (2016). Introduction: The geoeconomics and geopolitics of Chinese development and investment in Asia. *Eurasian Geography and Economics, 57*(3), 275–285.

Yeh, E. T., & Lewis, J. I. (2004). State power and the logic of reform in China's electricity sector. *Pacific Affairs, 77*(3), 437–465.

Yu, X., He, D., & Phousavanh, P. (2019). Transboundary environmental effects of hydropower: Sediment transport and geomorphology. In X. Yu, D. He, & P. Phousavanh (Eds.), *Balancing river health and hydropower requirements in the Lancang River Basin* (pp. 161–182). Singapore: Springer.

PART I

Hydropower and Resettlement Governance

CHAPTER 2

The Water-Energy Nexus of Southwest China's Rapid Hydropower Development: Challenges and Trade-Offs in the Interaction Between Hydropower Generation and Utilisation

Thomas Hennig and Darrin Magee

Introduction

China's rapid hydropower development is well known, and the territorial distribution of projects along key rivers is documented in national statistics (see, for instance, Li et al. 2018). However, a more sophisticated analysis of the spatio-temporal characteristics of this development remains

T. Hennig (✉)
Philipps-Universität Marburg, Marburg, Germany
e-mail: hennig@geo.uni-marburg.de

D. Magee
Hobart and William Smith Colleges, Geneva, NY, USA
e-mail: MAGEE@hws.edu

J.-F. Rousseau and S. Habich-Sobiegalla (eds.), *The Political Economy of Hydropower in Southwest China and Beyond*, International Political Economy Series, https://doi.org/10.1007/978-3-030-59361-2_2

elusive. In this chapter, we take a step towards filling this gap by examining Southwest China's hydropower development alongside its electricity utilisation, the so-called water-energy nexus. In the early 2000s, a handful of articles in English-language outlets (Dore and Yu 2004; Magee 2006) brought Yunnan's ambitious hydropower plans into the global spotlight. Since then, some studies have improved our knowledge of the status, drivers, and consequences of this development. However, most of the research has been qualitative and limited to only a few river sections in Yunnan, especially the Lancang-Mekong and to a lesser degree the Jinsha and Nu (Magee 2006; Magee and McDonald 2009; Hennig et al. 2013; Yu et al. 2019). Quantitative research is still rare; the few studies are limited and mainly focus on Yunnan's sections of the transnational Nu and Ayeyarwaddy basins (Kibler and Tullos 2013; Hennig 2016), both of which have sections among the world's greatest dam densities.

Southwest China has the world's largest ongoing hydropower development, whether measured by number of projects, installed capacity, affected river basins, or related infrastructure such as transmission lines. This build-out ranges from thousands of small projects to some of the world's largest. Our regional understanding of this development is still fragmented, as is our understanding of the local, regional, and long-distance utilisation of the electricity produced there. In this context, we address two aspects of the region's power sector: hydropower development itself, and the disposition of the huge amount of hydroelectricity produced in the Southwest. We are especially interested in the trade-offs that arise as one aspect of the water-energy nexus. Our data come from various yearbooks[1] and websites, as well as interviews and field visits conducted from 2010 to 2019. Our interpretation of these data is guided by more than a decade of intensive research in the area, much of which has been done in conjunction with Chinese partners, primarily at the Asian International Rivers Center in Kunming. This work has entailed many trips to hydropower stations, river basins, and prefectural offices.

Below, we first outline the general context of Southwest China's hydropower development. Then we focus in a more detailed and empirical way on key spatio-temporal characteristics of hydropower development

[1] All data in this article, if not cited otherwise, are based on China's National or Provincial Bureau of Statistics. Data were published in different years and in different sections of those national or provincial yearbooks. Some data were collected from printed versions, others from websites of statistical offices.

in three key provinces (Sichuan, Yunnan, and Tibet). In the second part, we shift the focus from hydropower development to its utilisation. This provides a better understanding of the water-energy nexus and the current mismatch between hydropower production and use. We differentiate between two scales. The export-scale analysis highlights the challenges related to long-distance transmission. We illustrate these challenges through the examples of Yunnan and Guangdong, which together comprise the world's largest sending and receiving areas for hydropower. Meanwhile, the local-scale analysis characterises the utilisation of hydropower in the local grid. Here we distinguish between power exports and local utilisation in power-intensive industries as well as aspects of rural electrification. We finish with a brief summary and conclusion.

China's Hydropower Development

General Context

Hydropower has been and will remain in the near future China's leading low-carbon power source. Hydropower's attractiveness to developers is due to its mature technology, its flexibility for serving base and peak load, its attractive economics (e.g. low to moderate tariffs), and its various other benefits (e.g. flood prevention, improved navigation and infrastructure, irrigation). In this planned economy with substantial public investment opportunities, and thanks to favourable natural endowments such as high mountains and big rivers, China's hydropower output has emerged as the world's largest. In 2019, China boasted 326 GW of conventional[2] installed hydropower capacity. By 2030, that figure is expected to rise to 410–460 GW (ERI and CNREC 2018).

China's hydropower resources are unevenly distributed. Outside the dry northern areas, many provinces have rather large hydropower potential. However, in many central and eastern provinces, hydropower cannot compete with rapidly growing power demand. Additionally, most of Eastern and Central China's technically feasible hydropower potential has been developed since the early 2000s. In comparison, China's southwestern provinces have an abundance of untapped hydropower resources.

[2] Hydropower without pumped storage. In 2019, China had another 31 GW of installed pumped storage capacity. If not mentioned separately, we always refer in this article to conventional hydropower.

Currently, the Southwest accounts for about 60% of China's hydropower production and there is still a great deal of undeveloped potential. For more than three decades, China's hydropower development has been gradually moving westwards, away from its major load centres and toward the Tibetan Plateau and adjoining mountain ranges. Various estimates (e.g. by the National Development and Reform Commission, NDRC) expect a hydropower expansion of 80–120 GW over the next decade for this area (ERI and CNREC 2018).

Like elsewhere, China's hydropower sector faces various challenges. The most important ones are fluctuating or seasonal production, stricter environmental restrictions based on growing awareness of hydropower's negative impacts, rising construction costs that limit its long-distance competitiveness (including tariffs and transmission costs), insufficient grids (congestion), competing demands for end uses of generated electricity, and massive oversupply in certain provinces. Despite these challenges, in most projections, China's hydroelectric sector will continue its growth well into the twenty-first century, when the technically feasible limit of China's hydropower potential will be reached (about 540 GW installed capacity, or 1750 TWh annual generation).

Southwest China Context

Southwest China is situated on the Tibetan Plateau and its adjoining mountain ranges, and accounts for the world's greatest spatial concentration of both potential and existing hydropower capacity. Most of the installed capacity has been added in the short period since 2000. However, hydropower implementation in the Southwest is unevenly distributed and can be unevenly characterised. On the one hand, the bulk of newly added capacity is based on projects in multi-dam cascades along a few major rivers. On the other hand, thousands of small and mid-sized projects have been implemented in a rather uncoordinated manner. Figure 2.1 shows large cascades in Southwest China (>1 GW). The Jinsha/Yangtze has emerged as the world's largest basin in terms of installed capacity and number of projects; it currently holds about 54% of China's installed hydropower capacity.

Hydropower development in Sichuan and Yunnan began to gradually increase in the late 1990s. Between 2011 and 2016, the installed capacity of the two provinces then doubled, increasing by more than 70 GW—almost equivalent to the entire hydropower capacity of either Canada or

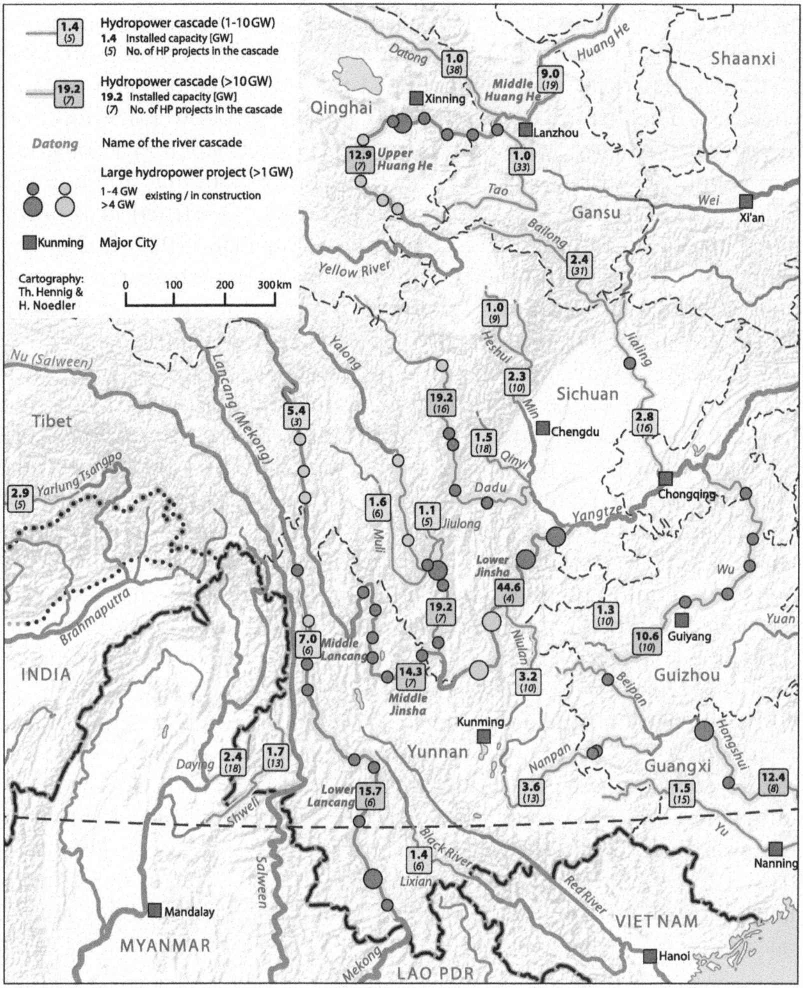

Fig. 2.1 Southwest China's dam cascades (>1 GW of installed capacity) and number of projects within the cascade

the United States. This dramatic rise is based on a rather small number of large projects and resulted in curtailment rates rapidly reaching 60 TWh in 2016 as electricity supply outstripped the residential demand of both provinces.[3] The first peak of Southwest China's hydropower development has thus been reached, with only a few large projects still under construction, mainly on various parts of the Jinsha. The Lancang, Yalong, and Dadu Rivers each had only one large project under construction as of the end of 2019. However, in China's next Five Year Plan (2021–2025), the number of projects will almost certainly rise again. In particular, Tibet will almost certainly face a dramatic rise in hydropower development pressure and could become the world's next hydropower hotspot.

Southwest China's hydropower production is highly seasonal, with over 70 per cent of the rainfall and correlated river discharge occurring during the rainy season (June–October). Annual storage regulation is still limited to a few projects along large rivers, namely the lower Jinsha and lower Lancang in Yunnan.[4] Otherwise, storage regulation for much of the region's installed hydropower capacity is rather poor, resulting in fluctuating seasonal generation.[5] Therefore, dry season procurement in the region still requires thermal power. Existing thermal plants may gradually be replaced by renewables; these renewables' production, however, would ideally be complementary to the seasonal fluctuations of hydropower, or else further curtailment would reduce effective utilisation.

Implementation of hydropower in the region is complex. Yunnan and Sichuan have 23 GW of combined small hydropower (SHP, 1–50 MW), a figure comparable to the installed capacity of the Three Gorges Dam. Southwest China has more than 10,000 hydropower plants; most of them are small. The peak of SHP construction was between 2005 and 2013, a few years earlier than large hydropower. Except on the Tibetan Plateau proper, virtually every feasible site in every watershed has already

[3] Curtailment refers to electricity that is not put to productive use through transmission to the grid or to a productive end-user such as a nearby factory.

[4] That is, the ability of dam reservoirs to store sufficient quantities of water over the year so as to ensure predictable and regular hydropower generation even through the dry season.

[5] Limited storage in reservoirs, of course, is generally a good thing for ecosystems, primarily because water quality in reservoirs tends to be lower than in free-flowing rivers, and because large reservoirs contribute to significant habitat fragmentation. Here, 'poor' is meant to qualify the dam system from an electricity production perspective only.

been developed. The governments of both provinces realised that SHP construction was so rapid and often uncontrolled that it resulted in a major oversupply of hydroelectricity, since local grids cannot absorb all the energy and transmission capacity to the main grid is limited (Hennig and Harlan 2018). As a result, in 2015 local officials instituted a ban on new small and mid-sized hydropower projects (under 250 MW in Yunnan and 300 MW in Sichuan) (Sichuan Daily 2016).

The complexity of large hydropower implementation in the region is exemplified by the Jinsha River, the Yangtze's major upstream tributary. The lower Jinsha constitutes the border between Yunnan and Sichuan. This is where China Yangtze Power (CYP), the owner of the Three Gorges Dam, is developing a four-project cascade with a capacity of 44.8 GW, more than double that of the Three Gorges Dam. Development rights actually belong to the provinces; however, since both provinces lack funds to develop the projects by themselves, they have taken partial ownership shares (15% each) in Jinsha YunChuan Hydropower, the development company constructing the dams, with the remainder owned by CYP. After completion, control of the dams will be fully handed over to CYP. Because of the interprovincial nature of these developments, decisions regarding utilisation of the cascade's energy are made by the NDRC at the central government level. Therefore, projects may not be shared equally between the two provinces or China's two large grids: State Grid Corporation of China (SGC), connecting Sichuan to the East, and China Southern Power Grid (CSPG), linking Yunnan with southern markets (Fig. 2.2). Xiangjiaba (6 GW), for instance, is entirely feeding into SGC, while Xiluodu (12.6 GW) is shared between the two provinces and grids. CYP sells electricity to receiving provinces (Guangdong, Zhejiang, and Shanghai) based on the benchmark coal power tariff in those provinces. This policy might be changed after 2020, when receiving provinces are allowed to adjust the tariff. Such a change would affect two forthcoming plants in particular. Wudongde (10.2 GW) will start power production in 2020, primarily feeding 8 GW into CSPG to serve Guangdong and Guangxi. The world's third-largest hydropower plant, Baihetan (16 GW), will start production in 2022–2023 and feed into SGC (to Jiangsu); its further utilisation is still under negotiation between the provinces and the central government.

The Jinsha therefore highlights the conflicts related to large-scale hydropower development. We explore these in further detail in Section 4. For now, it suffices to note that conflicts arise between developers like

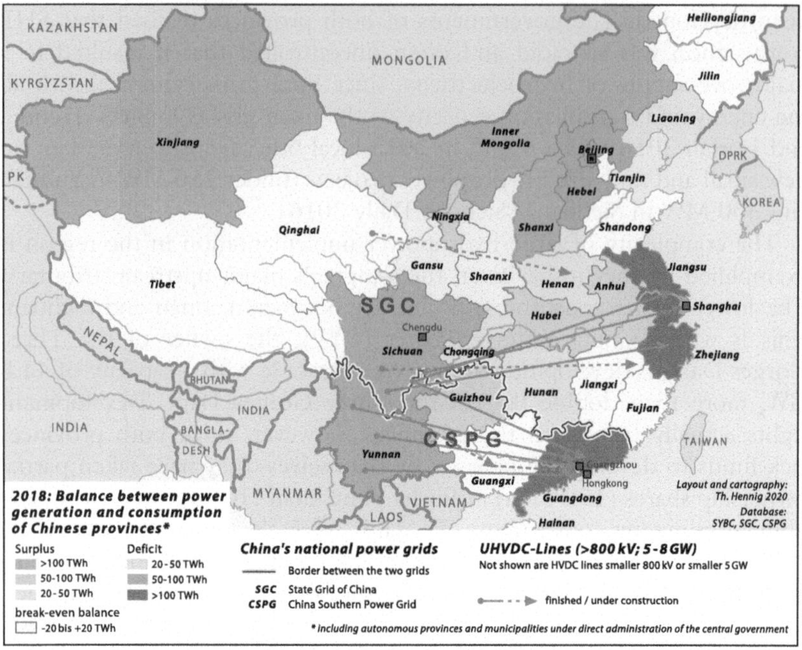

Fig. 2.2 China's major transmission corridors for hydropower (UHVDC lines > 5 GW and 800 kV) in the context of China's national power grids and its provincial electricity balance

state-owned enterprises, local provincial governments, governments in the demand area, and the central government. These disputes often result in delays and limitations of power evacuation and therefore large curtailments.[6] They also affect the planned generation portfolio of receiving provinces, especially the role of conventional thermal (coal and gas) and nuclear power. Disputes are intensified by the low tariff for hydro-exports, which is below the benchmark coal tariff in the sending area. While Yunnan or Sichuan as sending provinces have limited rights to argue for higher rates, which are set by the NDRC, they bear all the burdens related

[6] Power evacuation refers to the movement of power from a generator to the grid or load centre.

to implementation, such as resettlement requirements and environmental degradation (Liu et al. 2018).

Hydropower Development in Three Key Provinces

Sichuan

Sichuan holds China's largest provincial installed hydropower capacity—78.4 GW as of November 2019 (NEA 2019). Sichuan's hydropower development is almost exclusively located in the Yangtze basin, but there it is very unevenly distributed. Generally, along Sichuan's downstream portions of the basin, hydropower is all but ubiquitous, while upstream sections are often only locally developed and have a huge untapped potential. Just at the western descent of the Tibetan Plateau (between the Yalong River and Chengdu, Sichuan's capital), we identified more than 300 hydropower projects with installed capacities of 10 MW or more. Additionally, there is an unknown number of SHPs (1–10 MW) and larger run-of-river facilities in the basin itself.[7] The area has some of the highest hydropower densities and capacities in the world, which also means this part of the Yangtze basin is one of the world's most fragmented river systems. Sichuan's current and future hydropower development is gradually shifting to the upstream parts of the rivers and therefore to the inner parts of the Tibetan Plateau. Sichuan's Garze and Aba Prefectures, wherein lie the upper parts of the Jinsha and Yalong watersheds, still have almost 40 GW of unexploited hydropower potential (SGC 2019).[8]

We can further distinguish between individual sub-basins of Sichuan's Yangtze basin. Hydropower development in the Yalong (23.7 GW) and Dadu (25.6 GW) watersheds is most advanced, with the installed capacity of these two sub-basins ranking among the world's largest, comparable to those of the Indus, Yellow, or Pearl Rivers. However, more than 75% of this capacity stems from a handful of large projects along the two main rivers. While the Yalong primarily serves export markets (to Jiangsu

[7] Run-of-river hydropower plants are so named because they have limited to no reservoir storage, and their outflows equal their inflows.

[8] References to data on the websites of China's two large power grids, namely State Grid Corporation of China (SGC) and China Southern Power Grid (CSPG), refer specifically to the Chinese news section of their homepage. For example 'SGC 2019' means that the information was published in 2019 on the SGC page. Note that some older data (before 2018) can be found in the Social Responsibility Reports of SGC and CSPG.

and Chongqing), the Dadu's hydropower is mainly used within Sichuan (Ye et al. 2018). In contrast, other watersheds have a larger number of projects but are mostly of moderate size.

Yunnan

Compared to Sichuan, Yunnan's general hydropower development is much better known. In mid-2019, Yunnan's installed hydropower capacity was 64.5 GW (CSPG 2019). The province has more than 1900 grid-connected hydropower projects. In terms of capacity and number of projects, Yunnan's largest hydropower development is also found in the Yangtze basin. However, hydropower development is also advanced—albeit unevenly developed—in the other five transnational basins where Yunnan enjoys an upstream position. The lion's share of Yunnan's current and potential hydropower capacity is based on the cascades of its three large rivers: the Jinsha, Lancang-Mekong, and Nu-Salween.

The Lancang-Mekong is developed in three cascades. Except for one provincial project, all others belong to Hydrolancang (Huaneng), a state-owned utility. The lower cascade (15.7 GW), completed in 2012, has two reservoirs (Xiaowan and Nuozhadu) with annual storage. Electricity generated from this cascade serves various purposes, including power export along UHVDC and AC grid infrastructure as well as provincial utilisation (see Fig. 2.1).[9] The middle cascade (5.3 GW), is almost completed, with one final project under construction at Tuoba. Since 2018, this cascade has primarily served UHVDC transmission to eastern load centres; before that, most electricity was curtailed. The Lancang's upper cascade (1.9 GW in Yunnan and 8 GW in Tibet) was approved by Huaneng in 2019; it is still in the very early stages of development.

Yunnan's section of the Jinsha River is developed in two stages. The lower section (45 GW), shared with neighbouring Sichuan, was already described above. The upper section (13.8 GW) lies in Yunnan and is being developed by four different companies. This complex ownership structure is reflected in some rather uncoordinated management and utilisation patterns, serving four different UHVDC corridors. Huadian, its largest shareholder, plans an upstream extension (8 GW) in the Yunnan-Sichuan border area. This section faces strong opposition, especially for its

[9] Ultra-high-voltage DC (UHVDC) usually refers to direct current transmission lines capable of voltages greater than 500 kV.

potential impacts on the Three Parallel Rivers Protected Areas UNESCO World Heritage Site. Details of this planned cascade have not yet been made public.

Yunnan's third-largest river, the Nu-Salween, differs from the others. Alongside the transnational Amur-Heiliongjiang, it is frequently cited as China's last free-flowing main river, though it does have a very small number of dams on its principal headwaters in Tibet. Huadian holds most of the development rights for one cascade (23 GW), but the project was halted in 2004. Despite approval of a modified cascade in 2015, no project is currently under construction. The cascade still faces strong opposition, especially from environmental groups and scholars in China and overseas. However, of all of China's future hydropower cascades, the Nu has China's best potential cost-competitiveness in terms of construction and tariffs and this may result in its final approval being given with any new five-year plan. Such a decision would seriously damage and alter one of China's most unique ecosystems.

Beside these large cascades, Yunnan has an impressive 25 GW of additional installed hydropower capacity. This is split over the relevant watersheds of its three large rivers (e.g. along various Jinsha tributaries) and Yunnan's other three transnational basins (e.g. the Red River basin). All six basins have an additional installed hydropower capacity between two and six GW each. Our knowledge about this development and its various consequences is rather limited. The Ayeyarwaddy and Nu-Salween basins provide an apt illustration. Both rivers are among Asia's last (almost) free-flowing main rivers. In the English-language scholarly literature on dams, this area remains essentially *terra incognita*. Based on our research, we identified approximately 190 hydropower projects between 10 and 875 MW in the Chinese section of these watersheds. Additionally, we identified another 200 SHPs between 1 and 10 MW (Hennig and Magee 2017). In other words, in a region where our knowledge is limited, we find one of the world's largest hydropower densities. This causes a very high degree of river fragmentation and—due to the diversion structure of most hydropower projects—a dramatic seasonal dewatering of long river sections. This also affects biodiversity hotspots such as the Tongbiguan Nature Reserve. In these watersheds, there are no longer any free-flowing rivers or tributaries (Hennig 2016). Aside from environmental challenges, the high hydropower density also results in major utilisation and grid challenges stemming from the semi-autonomous status of most local grids. Despite the large number of projects, the connectivity of these local grids to Yunnan's main grid is limited and lacks sufficient seasonal power evacuation capacity.[10]

Tibet

Further west, Tibet's hydropower development is ready to take off. At the end of 2018, it had only 1.7 GW of installed hydro, with another 1.6 GW under construction. According to central government plans, that figure should rapidly grow to 24 GW in 2030 and a huge 110 GW in 2050. (Hu et al. 2015; ERI and CNREC 2018; Pang et al. 2018). The 110 GW figure is identical to Tibet's technically feasible hydro-potential, signalling planners' intent—and hubris—to wring every drop of technically feasible hydropower out of Tibet's rivers. Like its other Himalayan neighbours, China's planners, engineers, and central decision makers see Tibet not only as a crucial 'water tower' for Asia, but also a 'power tower'—the only difference is that the Chinese hydro development trajectory is markedly more dramatic than that of any other state. Meanwhile, Tibet's capacity for small hydropower (9 GW) is lower than in Sichuan and Yunnan. Unchecked development of Tibet's (small) hydropower resources without considering the lessons of similar development in Sichuan and Yunnan will without doubt bring ecological and social disruption in one of the world's most ecologically and culturally unique regions.

Tibet's future hydropower development will be shared by both Chinese grids, differing from Yunnan and Sichuan. Starting in 2015, Tibet started to export power, mainly hydroelectric and photovoltaic (PV), while at the same time importing hydroelectricity from Sichuan. In 2019, Huaneng approved the extension of its upper Lancang cascade (10 GW), which will serve China's Southern Grid, especially the Pearl River Delta. The construction costs per MW will be rather high (about double those of Yunnan's Xiaowan). The calculated power tariff for UHVDC exported from Tibet to Guangdong is also more than a third higher than exports from Yunnan—not surprising given the greater distances and challenging topography. This is higher than Guangdong's nuclear power tariffs, but cost-competitive with other alternatives like gas and offshore wind (MSR 2019). Higher tariffs mean developers will potentially recover their costs

[10] See *Local and regional utilisation of hydroelectricity* section below, as well as Hennig and Harlan (2018).

more quickly, and that electricity exports to Guangdong will play a relatively more significant role in Yunnan's provincial GDP figures, given royalty payments.

Utilisation of Hydroelectricity

In 2019, China produced about 1302 TWh of hydroelectricity. The world's largest dam (Three Gorges) contributes an immense 100 TWh, but this is merely 7.7% of China's total hydroelectricity output. One question motivating our work is the final disposition of this massive amount of hydroelectricity: What is it used for? Due to a paucity of data, we cannot conduct a detailed analysis. However, in seeking to discern general characteristics and utilisation trends, we generally distinguish between local utilisation within the grid (provincial or smaller administrative units) and distant utilisation, which we refer to as export. We illustrate this with Yunnan as a case study.

Export of Hydroelectricity

West-to-East (W2E) Transmission (Yunnan's Sending Perspective)
To understand the basic concept of hydropower export and its local utilisation, we must first understand the grid structure. As mentioned above, China's transmission and distribution network belongs to two state monopolies, SGC and CSPG.[11] Both national grids have their relevant local branches (provincial, prefectural, and county). Technically, however, only the national and provincial grids conduct large-scale, high-voltage transmission, while the local branches are in charge of local transmission and distribution (or utilisation). In other words, there is a big difference between these two scales. We use the CSPG as an illustrative example, focusing on the two provincial grids of Yunnan and Guangdong.

In mid-2019, Yunnan's total installed capacity reached 90.6 GW. This comprises 64.5 GW of hydro, with coal and renewables, especially wind power, accounting for the remaining 26.1 GW. In 2018, hydropower generated 83 per cent of Yunnan's electricity. About 65 mostly large hydropower projects feed directly into the national or provincial branch of CSPG. More than 200 other hydropower projects co-feed the provincial

[11] Inner Mongolia is partly served by a small third state grid monopoly that only fulfils local distribution.

and local grids. All these plants have to fulfil CSPG's high national quality standards (CSPG 2019). Yunnan's 1600 other hydropower projects feed into various local grids. These mostly small- to mid-sized plants serve rural electrification purposes. Changing to CSPG would mean costly investments to fulfil CSPG's technical security standards, an especially daunting task for older private hydropower stations. Therefore, many hydropower owners prefer to supply the local grids.

A substantial amount of hydropower generated on Yunnan's major rivers is sent to load centres elsewhere in Yunnan Province. The transmission is mainly within Yunnan's 500 kV AC ring corridor, e.g. the evacuation of hydroelectricity from Yunnan's Ayeyarwaddy or Red River basins. In the context of the middle Jinsha cascade, Yunnan's provincial transmission also includes China's first provincial UHVDC corridor to Wenshan. Both lines additionally support export to neighbouring provinces, especially Guangxi. Yunnan's power exports should be differentiated. In 2018, the entire export was 153 TWh (larger than Pakistan's power consumption). About 90% (138 TWh) falls under W2E transfer; the remainder is mainly AC export to neighbouring provinces or to neighbouring countries in Southeast Asia.

At a larger scale, in 2018, CSPG's complex interprovincial W2E transfer exported 218 TWh from various sources and provinces. Yunnan contributed almost two-thirds of that export, which came almost exclusively from hydropower. Yunnan's W2E export started in 1993 but scaled up only after 2009 with the commissioning of the world's first large UHVDC line (800 kV, 5 GW). Since 2014, electricity exports on this line have exceeded 100 TWh. Yunnan's transmission capacity is 31.3 GW, based on five DC lines originating from the Jinsha and Lancang dams and two 'back-to-back' DC converters collecting and transmitting electricity from all over Yunnan.[12] Currently, the world's first 800 kV bi-module UHVDC line is under construction in the province. From 2020 onwards, it will transmit power from the new Wudongde dam as well as from two Jinsha dams to Guangxi (3 GW) and Guangdong (5 GW).

Yunnan's hydropower exports primarily serve Guangdong, especially the Greater Bay Area (GBA), but also increasingly Guangxi and (since

[12] Back-to-back converters enable the interconnection of DC and AC grids, and of AC grids operating at different frequencies.

2019) Hainan.[13] Yunnan's long-distance hydro transmission has no equal worldwide. In fact, its immense size poses a major system-wide blackout risk for CSPG. To neutralise this threat, in 2016 China split Yunnan from CSPG's synchronised (unified) grid. The AC links to neighbouring Guangxi were disconnected, making Yunnan's grid once more a separate grid.

Yunnan's financial benefit from UHVDC exports is minimal. Therefore, it is looking for alternatives in a bid to increase its market share, such as AC exports to areas currently only served by China's State Grid. Yunnan is also looking to increase international exports to Southeast Asia, where it already has a long tradition of small-scale electricity exports. On the one hand, Yunnan imports hydroelectricity from two plants in Myanmar (Shweli and Dapeing), both of which feed directly into CSPG. On the other hand, Yunnan exports hydroelectricity to Vietnam (cumulative > 35 TWh as of 2018), but also to Laos and Myanmar. Yunnan is now in negotiations for a 3 GW HVDC line to power-hungry northern Vietnam. A comparable project linking Yunnan to Thailand is still being planned, but its completion does not appear realistic in the near future. Additionally, Yunnan is in discussions regarding two proposed 500 kV AC lines to Laos and Myanmar (see the chapter by Magee, this volume). Both should serve export and import of hydroelectricity, depending on seasonal and yearly variation.

West-to-East Transmission (Guangdong's Receiving Perspective)
Guangdong Province has the highest electricity consumption in China. Its 670 TWh demand made it the sixth-largest power consumer worldwide in 2019, behind China, the United States, India, Japan, and Russia. Its deficit alone (163 TWh) is larger than the total demand of many countries, leaving it heavily dependent on power imports. As an illustration, in 2019, the Chinese government announced its goal to reshape the Pearl River Delta and create the Greater Bay Area, which also includes Hong Kong and Macao. The new regional entity has a population of roughly 71 million. Its power demand at the end of 2018 was 93 GW, equating to 518 TWh of annual consumption (MSR 2019)—a figure roughly comparable to that of Germany. China's government estimates that by 2030 the GBA's power demand will rise to 146 GW, or 750 TWh annual

[13]The GBA includes the Pearl River Delta, Hong Kong, and Macao.

consumption. Further, the government plans to increase the GBA's low-carbon energy share from 54 to 80% by 2035, making it China's leading low-carbon generation area. This is expected to be achieved through a massive expansion of nuclear power, but also by an increase in gas imports, offshore wind, and especially hydropower imports. The GBA's power imports, mainly hydropower, are expected to rise from 40 GW in 2018 to 60 GW in 2035 (MSR 2019).

What do such figures mean for the grid, especially for our understanding of the trade-offs between hydro and grid developments? Since the early 1990s, Guangdong has been a net importer of electricity. In the early 2000s, CSPG's first HVDC lines were built to Guangdong, with capacities that appear small by today's standards. The first line (1.8 GW) is fed by three hydropower plants in the Nanpan basin (part of an upper Pearl River watershed) in the Yunnan-Guangxi-Guizhou border triangle. More HVDC-lines were subsequently built and W2E transmission increased substantially; in 2008, it exceeded 100 TWh for the first time. Except for a minor drop in 2010–2011, it has risen continuously (CSPG 2019). Currently, the GBA's import portfolio is 40 GW, including 23.5 GW of hydropower from Yunnan (three-quarters of Yunnan's exports), 3 GW from the Three Gorges Dam and another 2.1 GW from Datang's Longtan dam in Guizhou. However, it also includes coal power from Guizhou (8 GW) and from Hunan (1.8 GW).

West-to-East Transmission (Other Hydro-Fed UHVDC Corridors of China)

Like Yunnan, Sichuan also exports a major part of its annual > 300 TWh production of hydroelectricity. Its exports, based on the difference between generation and consumption, increased rapidly from 24 TWh in 2010 to 123 TWh in 2018.[14] The large rise is mainly attributable to the completion of Sichuan's three large UHVDC corridors connecting the province to the Yangtze River Delta. However, Sichuan's exports also rely on AC lines to neighbouring provinces, especially to Chongqing. The principal areas feeding Sichuan's electricity exports are the two Jinsha projects (to Shanghai) and the projects of the lower Yalong (to Jiangsu and Chongqing). The existing lines cannot absorb all the hydropower, so Sichuan's curtailment increased to more than 30 TWh in 2017 (Qiang

[14] Comparable statistical data from China's State Grid differ from these figures, being 3 TWh lower in 2010 and 17 TWh higher in 2017.

et al. 2019). Half of Sichuan's curtailment is in the Dadu watershed; hydropower projects in the Tibetan Plateau also suffer from oversupply (Ye et al. 2018). Therefore, in 2019, State Grid started construction of Sichuan's fourth UHVDC line (8 GW), which will significantly reduce the region's curtailment. This line will primarily take hydroelectricity from the Yalong watershed and transmit it to Jiangxi and Zhejiang. Yet another 8 GW line, set to begin operation in 2021, has been proposed from the Baihetan dam on the Jinsha to Jiangsu. However, a construction timeline for this line has not yet been announced.

Further downstream, the Three Gorges Dam and the neighbouring Gezhouba Dam export a major portion of their electricity via one 1.8 GW and four 3 GW HVDC lines. Beside the lines to Guangdong, others transmit to Shanghai and Jiangsu. As a result, the Yangtze River Delta's hydroelectricity imports have become the world's second-largest. Finally, China's fourth-largest hydropower export corridor is currently under construction. From mid-2020, it will export 8 GW of hydroelectricity from the upper Huang (Yellow) River in Qinghai to Henan, displacing some of Henan's ageing thermal power.

Local and Regional Utilisation of Hydroelectricity

In 2018, Yunnan exported half of its electricity and used the other half within the province (156 TWh). This figure is comparable with the power demand of Asian neighbours, e.g. Vietnam, Thailand, or Malaysia. However, in terms of GDP, Yunnan is one of China's poorest provinces, prompting us to ask why this land-locked, peripheral, poor province needs so much electricity. Below we characterise the utilisation of hydropower within CSPG, focusing especially on Yunnan.

Yunnan's industrial sector is responsible for about three-quarters of all electricity consumption in the province. In contrast, residential demand is only about 15% and the tertiary sector demands only 11%. About 80% of Yunnan's industrial demand is used in power-intensive industries (e.g. electrolytic aluminium, silicon purification, yellow phosphorous, ferroalloy processing). This level of dependency is larger than in many other provinces, and much larger than in developed countries. Between 2013 and 2017, Yunnan's power demand unexpectedly stagnated, and only after 2018 did it start to rise again. While residential demand has continuously risen, demand in power-intensive industries has stagnated or locally even decreased. Yunnan's peripheral location, relatively high tariffs (set by NDRC), and older technology have forced many industrial end-users to close temporarily. Such closures additionally contribute to Yunnan's large hydropower curtailment rates.

Aluminium production is one of the most electricity-intensive industries. In many countries, large hydropower plants have been built to supply (as captive power) 'green' and rather cheap electricity for nearby aluminium smelters. The situation in China, which in 2018 produced 56.7% of the world's primary aluminium, is different. A 2019 report by the German Energy Agency (DENA) notes that China's 217 aluminium smelters consumed 775 TWh (Willuhn 2019). Unlike many other countries, most of the smelters' power supply comes from coal. Hydropower-rich provinces in the Southwest currently produce only 14% of China's aluminium. However, the central government has recently begun pushing smelting capacity from China's leading provinces (Shandong, Henan, and Shanxi) to Yunnan in order to replace coal-powered smelters with hydropowered ones. After finishing construction between 2019 and 2021, Yunnan will host two of China's largest smelters and will become one of the world's largest hydropower-based producers of aluminium (labelled as 'green' aluminium). Yunnan will more than double its smelting capacity and face serious environmental constraints, especially regarding alumina refinery residues and emissions. With this development, some of the world's largest aluminium producers—Chinalco, Hongqiao, and Shenhuo—will enter Yunnan, either by merging with local producers (Chinalco) or by constructing new plants (Hongqiao and Shenhuo). The largest beneficiary from this restructuring will be Wenshan Prefecture of southeastern Yunnan, which lies in the Nanpan and Red River watersheds. Northwestern Yunnan will also benefit on a smaller scale. In this area, initial plans for a large smelter were cancelled due to serious environmental concerns about potential impacts to Lijiang, a UNESCO World Heritage city, and the Three Parallel Rivers area. Wenshan's local hydropower supply is not able to provide sufficient electricity, so the smelters will benefit from imports from elsewhere in the province. This includes 500 kV transmission corridors and China's first provincial UHVDC line transmitting hydroelectricity from the middle Jinsha (especially Guanyinyan) to Wenshan. From 2020 onward, Yunnan's provincial power demand will likely rise substantially.

Yunnan's aluminium production and grid integration differs from other energy-intensive industries like silicon production. In western Yunnan, especially prefectures in the Ayeyarwaddy and Nu basins, local grids have limited transmission capacity to Yunnan's provincial grid, so silicon smelters rely on the huge seasonal surplus of hydroelectricity within the local service area. Production of purified silicon is Yunnan's second-largest industrial electricity consumer. The detailed interactions between hydro-generation and silicon smelters are described by Hennig and Harlan (2018). One side effect of smelters' dependence on cheap hydropower, though, is that they must contend with seasonal fluctuations in hydropower output and tariffs. Therefore, they mainly compete with Xinjiang's cheap and rapidly growing coal-powered silicon processing industry.

Another emerging power-intensive industry using western Yunnan's huge power surplus is data centres and networks. Only the American tech industry utilises more electricity than China's tech giants; in 2018, about two-thirds of power for Chinese tech came from coal. As with aluminium, there is a growing trend to use green power, especially hydro. In 2017, Alibaba built one of its data centres in western Yunnan's Baoshan Prefecture in order to tap into local hydropower.

Rural Electrification

A third destination for hydropower is Yunnan's (and China's more generally) impressive and successful rural electrification programme. As of 2018, Yunnan had 12.1 GW of rural hydropower (as opposed to SHP), which produced roughly 43 TWh of electricity annually. This figure alone is double the current hydropower production of Laos, which is often called the battery of Southeast Asia. China's government first promoted SHP for rural electrification in the 1950s. In 1982, Beijing intensified those efforts by establishing 'rural electrification counties' in southwestern China (see the chapters by Harlan and Ptak). This policy gave local governments the authority to approve, operate, and re-invest revenues from SHP. Thousands of very small plants were constructed near townships and became the basis for the first local grids. With economic liberalisation and rapidly growing power demand, the central and provincial governments provided funding for SHP construction, and smaller dams were gradually replaced by larger projects, resulting in 2 GW of rural hydropower in Yunnan by 2000. Later, additional programmes such as 'SHP replaces firewood' and the Kyoto Protocol's Clean Development Mechanism supported a massive expansion of rural hydropower. Yunnan's prefectures became entirely electrified between 1997 (Dehong)[15] and 2012 (Nujiang), a process based primarily on rural hydropower as in many other parts of China (Hennig and Harlan 2018).

Yunnan's massive rural hydro boom occurred within a decade (2005–2015) and was primarily driven by private investment from southeastern and eastern China, where a similar boom happened a decade earlier. The ensuing development of nearly all potential sites in Yunnan caused a massive oversupply. As noted above, most plants only feed into the local grids, meaning that the hydroelectricity must be used locally. However, supply often exceeds local demand. For long periods in the rainy season, many hydropower projects run idle and cannot completely feed into the grid, despite contract arrangements with Yunnan Power Grid (YPG).[16] In 2016, Yunnan's government halted new SHP construction as Sichuan had done. This policy change has affected the local atmosphere: before 2016 local governments praised SHP development, but now they are more circumspect.

Summary and Recommendations

Southwest China is currently the world's largest hydropower hotspot in terms of capacity and number of projects. While hydropower development is already advanced in Sichuan and Yunnan, it is still in its initial stages in Tibet, where growth will likely be most dramatic and highly problematic. Southwest China's spectrum of hydropower projects ranges from small to the world's largest. Extremely rapid development, especially over the past decade, was to some extent out of the provinces' control and resulted in substantial overcapacity and curtailment. Larger projects (cascades) were mainly driven by the central government, while thousands of smaller projects were mainly implemented by local governments. Serious environmental problems ensued (e.g. river fragmentation, flow regulation and large-scale diversion, and dewatering of rivers and watersheds). Uncontrolled hydro development of entire watersheds which lacked coordination with local grid and industrial development resulted in a halt, likely temporary, in new small- and mid-sized dam construction.

[15] Dehong was the second prefecture in China to be completely electrified.

[16] For a more comprehensive discussion of hydropower's under-utilisation in the southwest, see also Magee's chapter in this volume.

Due to this policy shift, in both Yunnan and Sichuan (but not in other parts of China), small hydropower is no longer assumed to be a green and sustainable energy source. As other countries and Chinese provinces around the Tibetan Plateau revise their hydropower development plans, it is critical that they not only assess the merits and impacts of individual projects, but also the cumulative impacts of multiple projects in the same basin.

At the local scale, Southwest China's overdevelopment suffers from insufficient local demand and weak local grids, plus a lack of provincial power evacuation corridors. It is even more serious at a broader scale, affecting the large cascades along Yunnan's and Sichuan's major rivers. Disputes about hydropower utilisation among governments (central and provincial) and other stakeholders (e.g. developers, grid, and large users) will have far-reaching consequences. These include the relocation of power-intensive industries, the transfer of hydropower to high-demand areas, and deciding whether to boost autonomous power production in high-demand areas (e.g. replacing coal by nuclear or gas) or increase dependency on hydropower imports. Negative outcomes like the large hydropower oversupply in Yunnan and Sichuan and the acute negative environmental and social consequences in the dams' vicinity have been the direct result of such disputes.

As China's rapid and impressive hydropower development continues, the need becomes increasingly dire for transparent decision making, comprehensive impact assessments, and realistic need analyses about power production and economic development. The negative ecological and social impacts of hydropower development are unquestionably acute at the local scale and have historically been undervalued and oversimplified. Research on such impacts has progressed substantially since the early days of Southwest China's hydropower boom, yet development decisions continue to be driven by power sector planners, hydropower developers, and local governments; these actors, despite some recent trepidation about unfettered SHP projects, have often overestimated the benefits of hydropower and downplayed its risks. Southwestern China's large river systems and steep mountains make the region particularly well-suited for hydropower development, a situation that bodes favourably for the region's continued and increased role as a renewable energy battery for a wide swath of southern and southeastern China and mainland Southeast Asia. Yet in light of the region's rich ecological and cultural heritage, and of hydropower's potential to make enduring and irreversible alterations to physical and human landscapes, a more measured and thoughtful approach that recognises the non-hydropower values inherent in every river is warranted.

Acknowledgements Thomas Hennig acknowledges funding from DFG (German Research Foundation), grant HE 5951, 4-1 and 6-1.

References

China Southern Power Grid (CSPG). (2019). *2019 company headlines*. http://www.csg.cn/xwzx/2019/gsyw/. Accessed 3 June 2020.

Dore, J., & Yu, X. (2004). *Yunnan hydropower expansion: Update on China's energy industry reforms an the Nu, Lancang & Jinsha hydropower dams* (p. 38). Chiang Mai and Kunming: Chiang Mai University and Green Watershed.

Energy Research Institute (ERI) and China National Renewable Energy Centre (CNREC). (2018). *China renewable energy outlook*. http://boostre.cnrec.org.cn/index.php/2018/11/27/china-renewable-energy-outlook-2018/?lang=en. Accessed 3 June 2020.

Hennig, T. (2016). Damming the transnational Ayeyarwady basin. Hydropower and the water-energy nexus. *Renewable and Sustainable Energy Reviews, 65*, 1232–1246.

Hennig, T., & Harlan, T. (2018). Shades of green energy: Geographies of small hydropower in Yunnan, China and the challenges of over-development. *Global Environmental Change, 49*, 116–128.

Hennig, T., & Magee, D. (2017). Comment on 'An index-based framework for assessing patterns and trends in river fragmentation and flow regulation by global dams at multiple scales'. *Environmental Research Letters, 12*(3), 038001.

Hennig, T., Wang, W., Feng, Y., Ou, X., & He, D. (2013). Review of Yunnan's hydropower development. Comparing small and large hydropower projects regarding their environmental implications and socio-economic consequences. *Renewable and Sustainable Energy Reviews, 27*, 585–595.

Hu, Y., Huang, W., Chen, S., Wang, J., & Liu, Y. (2015). Analysis of the hydropower generation cost and the affordability of the hydropower on-grid price in Tibet. *Journal of Renewable and Sustainable Energy, 7*(1), 013131.

Kibler, K. M., & Tullos, D. D. (2013). Cumulative biophysical impact of small and large hydropower development in Nu River, China. *Water Resources Research, 49*(6), 3104–3118.

Li, X. Z., Chen, Z. J., Fan, X. C., & Cheng, Z. J. (2018). Hydropower development situation and prospects in China. *Renewable and Sustainable Energy Reviews, 82,* 232–239.

Liu, B., Liao, S., Cheng, C., Chen, F., & Li, W. (2018). Hydropower curtailment in Yunnan Province, southwestern China: Constraint analysis and suggestions. *Renewable Energy, 121,* 700–711.

Magee, D. (2006). Powershed politics: Yunnan hydropower under great western development. *China Quarterly, 185,* 23–41.

Magee, D., & McDonald, K. (2009). Beyond Three Gorges: Nu River hydropower and energy decision politics in China. *Asian Geographer, 25*(1–2), 39–60.

Morgan Stanley Research (MSR). (2019). *China's urbanization 2.0: Beneficiaries of the greater bay area's transition to low-carbon energy.*

National Energy Administration (NEA). (2019, December 27). *Sichuan hydropower exports surpass 1,000 TWh.* http://www.nea.gov.cn/2019-12/27/c_138661327.htm. Accessed 3 June 2020.

Pang, M., Zhang, L., Bahaj, A. S., Xu, K., Hao, Y., & Wang, C. (2018). Small hydropower development in Tibet: Insight from a survey in Nagqu prefecture. *Renewable and Sustainable Energy Reviews, 81,* 3032–3040.

Qiang, Y., Qian, C., Yi, L., Quan, T., Zhichao, R., & Ruiguang, M. (2019). Study on the situation and countermeasures of power supply & demand in Sichuan province under the background of the supply-side reform. *Advances in Economics, Business and Management Research, 66,* 17–21.

Sichuan Daily. (2016). *Sichuan: 13th Five-Year Plan brings halt to small hydropower under 50 MW.* http://www.gov.cn/xinwen/2016-10/30/content_5126079.htm. Accessed 3 June 2020.

State Grid Corporation of China (SGC). (2019). *Headline centre.* http://www.sgcc.com.cn/html/sgcc_main/col2017021152/column_2017021152_1.shtml. Accessed 3 June 2020.

Willuhn, M. (2019). China's aluminum smelters alone could shift 2.3% of daily electricity demand. *PV Magazine,* https://www.pv-magazine.com/2019/09/05/chinas-aluminum-smelters-alone-could-shift-2-3-of-daily-electricity-demand/. Accessed 3 June 2020.

Ye, Y., Huang, W., Ma, G., Wang, J., Liu, Y., & Hu, Y. (2018). Cause analysis and policy options for the surplus hydropower in southwest China based on quantification. *Journal of Renewable and Sustainable Energy, 10*(1), 015908.

Yu, X., He, D., & Phousavanh, P. (2019). *Balancing river health and hydropower requirements in the Lancang River basin.* Singapore: Springer.

CHAPTER 3

Leaving the Three Gorges After Resettlement: Who Left, Why Did They Leave, and Where Did They Go?

Brooke Wilmsen, Andrew van Hulten, and Yuefang Duan

Introduction

Large dams generate profound effects, perhaps most acutely felt by the people forced to make way for them. It has been estimated that between 40 and 80 million people have been displaced by dams around the world (WCD 2000; Scudder 2005). Large dams tend to displace people who are already poor, uneducated, and politically marginalised (Scudder 2005).

Since the 1980s, multilateral institutions have promulgated best practices in resettlement standards to mitigate the effects of their projects on local populations (Vanclay 2017). Meanwhile, national and sub-national

B. Wilmsen (✉) · A. van Hulten
La Trobe University, Melbourne, Australia
e-mail: B.wilmsen@latrobe.edu.au

Y. Duan
China Three Gorges University, Yichang, China

J.-F. Rousseau and S. Habich-Sobiegalla (eds.), *The Political Economy of Hydropower in Southwest China and Beyond*, International Political Economy Series, https://doi.org/10.1007/978-3-030-59361-2_3

governments have implemented their own laws, regulations, and policies to ensure displaced people are given the support they need to rebuild livelihoods and communities (Vanclay 2017). Over the same period, the resettlement industry has evolved to implement these standards, regulations, and laws at the project level. However, whilst these efforts have produced tangible improvements in resettlement practices (Cernea 2016; Vanclay 2017), involuntary resettlement continues to cause impoverishment (Reddy et al. 2015; Smyth et al. 2015; Wilmsen et al. 2019).

China is home to more than half of the world's large dams (Wilmsen et al. 2011a). Dams have been an integral component of the Chinese Communist Party's (CCP) plans to modernise the country. Since market-based reforms began in the late 1970s, China's hydropower sector has grown 20-fold to an installed capacity of 352 GW (IHA 2019). China now accounts for over a quarter of the entire world's hydroelectricity generation capacity.

This monumental construction effort has displaced villages, towns, and entire cities across China. Dam-induced involuntary resettlement has had profound—and often negative—effects on the lives of tens of millions of Chinese people. In recent years, the Chinese government has recognised the historical injustices caused by various resettlement projects and has started providing retrospective compensation and development assistance to some affected communities (Wilmsen 2016a). Since the 1990s, China has gradually improved its involuntary resettlement laws, regulations, and policies, and evidence is emerging that these efforts are improving resettlement practices and outcomes (Cernea 2016).

In the coming decades, China's resettlement theories, policies, and practices will be of increasing international importance. Since the late 1990s, the Chinese government has encouraged Chinese firms to expand into international markets, first under the Go Out Policy and more recently through its Belt and Road Initiative (BRI) (Hensengerth 2013; Han 2018). Meanwhile, China is becoming a 'norm maker' as well as a 'norm taker', increasingly asserting its voice in existing international fora and creating new multilateral institutions, like the Asian Infrastructure Investment Bank (AIIB) and Silk Road Fund (Hensengerth 2013; Urban et al. 2017; Wilmsen et al. 2020). In this regard, China's National Research Centre for Resettlement has played an increasingly important role in shaping the intellectual discourse at the international level and training practitioners and scholars in the Chinese way of carrying out resettlement projects (Rogers and Wilmsen 2020).

Recognising the growing importance of Chinese resettlement policies and practices, this chapter provides an overview of our research, spanning nearly two decades, on the resettlement practices and outcomes for the Three Gorges Dam resettlement. We begin by discussing the unique political and economic circumstances of the project. We then describe the resettlement outcomes of two counties involved in the project. In the third section, we present a new analysis of those who disappeared from our sample and the resettlement area. In doing so, we begin to address an important gap in our own research and the resettlement literature more broadly—namely, the degree to which households leave resettlement communities either because they are unable to rebuild sustainable livelihoods or to take advantage of better opportunities elsewhere. We conclude by reviewing the existing evidence on resettlement in the Three Gorges Dam area and identify areas for future research.

The Three Gorges Dam: An Important But Unique Case

The Three Gorges Dam is the world's largest dam. Since the founding of the PRC, the Three Gorges Dam was seen by the CCP as a critical nation-building project that would, amongst other things, provide a sustainable source of energy to drive industrialisation, facilitate navigation and water management in central China, and showcase China's emergence as an economic and technological superpower. In the 1990s, the project was the subject of extensive criticism. Concerns were raised about its technical and economic feasibility, the environmental risk it posed, and its effect on the 1.3–1.4 million people it displaced, 87% of whom were rural people (Heggelund 2004; Xinhua 2016). The sheer scale and political sensitivity of the project—in particular, the possibility it might generate political unrest—meant the central government was far more attentive to the resettlement than it had been for previous dam projects.

The Chinese central government intervened at crucial junctures to address the hardship inflicted on resettled communities. In 1993 and 2001, it announced regulatory and economic development measures to promote local economic development. Two innovative programmes are worth mentioning here. The Partnership Support Programme (PSP) encouraged the eastern provinces to partner with counties in the Three

Gorges region to generate employment and promote economic development (Wilmsen et al. 2011a). By 2003, 700 primary and middle schools and 2500 industrial projects were constructed under the PSP, resulting in 16 billion Renminbi (RMB) worth of investment in the Three Gorges region (McDonald et al. 2008). The Development Assistance Fund (DAF) was set up to re-invest earnings from power generation into the local economy. The funds are distributed between Chongqing and Hubei according to the number of resettlers in each province. The DAF's effectiveness in promoting economic development, however, remains an open question (see Wilmsen 2016a). In 2011, the State Council intervened again by publicly acknowledging the hardships experienced by resettlers—including reduced arable farmland, environmental degradation, unemployment, low levels of education, the failure of small businesses, and inadequate social security payments—and announced a raft of new development interventions (Duan and Wilmsen 2012). Just as important in shaping resettlement in the Three Gorges has been the high level of economic growth seen in China, in large part a result of China's investment-heavy, state-led economic development mode. It is important to keep in mind these unique economic and political circumstances when thinking about how the lessons of the Three Gorges resettlement might be applied elsewhere.

The Three Gorges Dam has been a central player in several important milestones in the evolution of China's reservoir resettlement laws and policies. In the 1990s, the World Bank and other multilaterals began developing policies and standards of best practice to improve resettlement outcomes. As part of these efforts, resettlement was re-conceptualised as a development opportunity that required the full participation of affected communities. Whilst many member countries were sceptical about this new approach, China quickly embraced it as official government policy. In fact, the Three Gorges resettlement project represents China's first attempt to implement the notion of Resettlement with Development (RwD), a distinctly Chinese approach to planning resettlement projects (McDonald et al. 2008). The notion of RwD has been widely criticised, most broadly for facilitating and justifying the displacement of marginalised people in the name of economic development (Rogers and Wilmsen 2020).

The Three Gorges case plays an important role in another shift in Chinese reservoir resettlement policy, namely China's gradual shift away from land-based resettlement practices towards a policy of relocating rural people to urban areas. Chinese regulations for hydropower resettlement still require the government to provide displaced rural households with cash compensation and new land upon which to re-establish their livelihoods (State Council 2006).[1] The early resettlement plans for the Three Gorges Dam project complied with this law. The initial plan—embedded in the aforementioned 1993 regulations—required unused land in the hills and mountains around the reservoir to be cleared and made available to resettled rural households. However, devasting floods in 1998, which took the lives of some 3000 people, required a drastic revision of this strategy. The government concluded the floods were exacerbated by deforestation of the hills and mountains along the Yangtze River and, through the National Forest Conservation Programme (NFCP), curtailed further land clearing and encouraged re-forestation (Heggelund 2004). This restricted the supply of land available to resettled farmers. Responding to this, new regulations announced in 2001 encouraged displaced farmers to be moved out of the area and sometimes off the land altogether. Around 190,000 rural residents (15% of the total) were resettled in 11 provinces outside the Three Gorges area (Xu et al. 2013). Approximately, 200,000 rural residents were moved into nearby towns within the area (Wu and Liao 1999).

Although the floods in 1998 were the proximate cause for the shift in the Three Gorges resettlement plan, it mirrored the Chinese government's increasing appetite for urbanisation more broadly. Over the past decades, the Chinese government has gradually loosened limits on internal migration. In 2014, the State Council announced its most proactive commitment to promoting urbanisation in China, the National New-Type Urbanisation Plan (2014–2020). Under this plan, there is an aim to increase the number of urban residents by 100 million by 2020 (Tang 2016). This urbanisation target will be met through a combination of land reforms, urban renewal projects, and resettlement programmes.

Reservoir resettlement plays a role in these plans (Chen 2016; Wilmsen 2018a). Although the law governing reservoir resettlement in China still mandates a preference for land-based resettlement, in 2019 the new Land Administration Law actively promoted the resettlement of rural people

[1] See Article XIII of the Regulations for Resettlement for Large and Medium Hydropower and Water Conservation Projects. For more on resettlement policies, see the chapters by Tilt and Chen as well as Habich-Sobiegalla and Plümmer, this volume.

to urban areas (Duan et al. 2019). Cernea (2016: xliii) calls urbanisation resettlement 'a new chapter in the history of resettlement programs managed by the State'. Whilst the urbanisation drive came after resettlement of the Three Gorges Dam region was complete, changes to the resettlement plan for the Three Gorges Dam project after the 1998 floods provide an early test-case of what happens to rural people who are resettled to cities.

In the next section, we provide a brief overview of our research on long-term resettlement outcomes in the Three Gorges. When reviewing this evidence, it is important to remember the Three Gorges Dam is an important but unique case. Its sheer scale and political sensitivity—in particular, the possibility that it would generate widespread political unrest—meant the central government was far more attentive to the resettlement than may otherwise have been the case. As such, the resettlement outcomes observed in the Three Gorges are not representative of other reservoir resettlement projects carried out in China.

Resettlement Outcomes at the Three Gorges Dam

Although our work in the Three Gorges began in the 1990s, our first large-scale study was carried out in 2004. Given the perennial difficulty of establishing a reliable control group (Wilmsen and van Hulten 2017), six distinct study populations were randomly sampled based on resettlement location (Zigui County and Badong County) and type (rural-to-rural, rural-to-urban, urban-to-urban). We surveyed 521 households across these six cohorts. By this time, the households were already living in their new residences, and the rural-to-rural cohort had been provided with new farmland. The survey asked displaced households to reflect on their livelihoods before and after resettlement (in 2003). Semi-structured interviews were also conducted with a selection of resettled households, government officials (county, provincial, and national), and local enterprise managers.

The sustainable livelihoods approach (SLA) was used to devise questions and analyse the subsequent data. The SLA provides a systematic understanding of the factors that constrain or enhance livelihood opportunities, including livelihood assets, policy and institutional processes, livelihood strategies, and vulnerabilities (Scoones 2009). By analysing these factors, we sought to draw conclusions about the sustainability of post-resettlement livelihoods (DFID 1999).[2]

Overall, our analysis painted an unfavourable picture of resettlement outcomes in the Three Gorges. Despite improvements to infrastructure and improved housing, incomes generally declined, livelihoods were dismantled, and permanent jobs were replaced by more precarious forms of employment (McDonald 2006; Wilmsen et al. 2011a, b). As a result, many households were drawing down their personal savings and borrowing money from relatives and friends in order to cover their daily expenses (Wilmsen et al. 2011b). In short, despite the government's attempts to provide compensation and share the economic benefits of the project through a range of development initiatives, most of the sampled households reported lower living standards compared to before their resettlement (Wilmsen et al. 2011a).

There were also changes in the type of work people were undertaking following resettlement. Employment in agriculture decreased significantly throughout the study area, whilst employment in construction, small business, and transport increased. Many low-skilled households, including those forced off the land, found work in construction. This work was short-term, insecure, and associated with lower average household incomes, raising concerns about the viability of many households' livelihood strategies. Other households reported younger family members leaving to find work in nearby cities, a common livelihood strategy in China. This resulted in a hollowing out of the resettlement communities, with farming work done disproportionately by older people and women with children (Wilmsen 2018b).

The Chinese government hoped the Three Gorges Dam and associated employment and education initiatives would promote inclusive economic growth in the region. Although many households were able to find jobs, pay and security were often low, and there was little evidence of long-term capacity building—that is, of resettled people being equipped with

[2] In the 1990s, SLA emerged from the academic literature on participatory development and was used extensively by development agencies to plan and assess development projects. More recently, it has been the subject of extensive critique (Scoones 2009; Reddy et al. 2015; Smyth and Vanclay 2017), in particular for the complexity of translating the framework into practice at the project level. Nonetheless, at the time of our research it was widely accepted and, despite its shortcomings, provided a useful framework for framing our research questions in the Three Gorges. For more details on our research methods and limitations, see Wilmsen (2016a) and Wilmsen and van Hulten (2017).

the skills, work experience, and qualifications they needed to access the higher-paid jobs being created in the region. Indeed, our interviews revealed enterprise managers had prejudiced attitudes about employing ex-farmers, who were seen as unskilled and poorly educated.

Reflecting this, our cohort of rural-to-urban resettlers had the worst overall resettlement outcomes. For example, they were more likely than rural-to-rural and urban-to-urban resettlers to report insufficient income to cover daily expenditures, inadequate access to food, and high-levels of debt (Wilmsen et al. 2011a). Not only did these (mostly) ex-farmers have great difficulty finding work in urban labour markets, the financial compensation they were provided was often insufficient to re-establish new residences whilst also paying for basic needs such as food, which they had previously supplied for themselves.

Our findings mirror those of other studies conducted in the Three Gorges in the early 2000s (Jun 2000; Duan and Steil 2003; Tan et al. 2005; Hwang et al. 2007). Indeed, research commissioned by the Chinese government itself revealed inadequate compensation, high unemployment, ineffective training programmes, and insufficient land to resettle displaced farmers (Duan and Wilmsen 2012; Tilt 2014).

In short, it appeared that the unhappy history of involuntary resettlement in China was repeating. We were mindful, however, that these studies, including our own, were carried out at a time of great flux within communities and the regional economy. Our research also revealed that *some* households—in particular, those who had secured employment in an enterprise and farmers with sizeable land holdings—were re-establishing their livelihoods. Meanwhile, there was a growing recognition in the resettlement literature of the need for research on long-term resettlement outcomes (Scudder 1997; Cernea 1999).

With this in mind, we returned to our original field sites in 2012 to carry out another round of surveys and interviews. We found 351 households to respond to the follow-up survey, 67% of the original sample. Using the same suite of questions as the original survey, we measured changes in a range of important livelihood measures between 2003 and 2011. The data revealed dramatic improvements across a range of variables. Across both counties and all three resettlement categories, average real incomes surged past their pre-resettlement levels, exhibiting significant catch-up with the urban incomes of the nearest large city, Yichang, which we used for comparison purposes (see Wilmsen 2016a; Wilmsen and van Hulten 2017). There was a decline in the proportion of households reporting they did not have enough food, a decline in income inequality, and an improvement in self-reported social well-being. Overall, the results of our follow-up survey were both surprising and heartening, given the unhappy situation we observed in 2004. Our results demonstrate it takes time—perhaps many years—for economies and communities to recover from the disruption caused by large infrastructure construction and resettlement.

We must be careful, however, about drawing optimistic conclusions about Chinese reservoir resettlement practices from these results. As discussed previously, the Three Gorges Dam is a unique case by virtue of its sheer size, its status as a nation-building project, and the amount of international attention it received. Further, despite the dramatic improvements in livelihoods we observed, a significant proportion of households still reported income and food insecurity in 2011. Again, the outcomes for the cohort of rural-to-urban resettlers were particularly concerning. Although the real incomes of rural-to-urban resettlers grew over 400% between 2003 and 2011, their average incomes were still the lowest of the three resettlement categories (Wilmsen 2018a). Rural-to-urban households were also less likely to have secured jobs in enterprises and more likely to be employed in insecure and menial jobs compared to their urban-to-urban counterparts. Of the three resettlement categories, rural-to-urban migrants were the most likely to report they did not have enough food and that their incomes were insufficient to cover their daily expenses. In contrast, the rural-to-rural households were doing quite well. Despite the low quality of their land following resettlement, the majority of rural households in both Badong and Zigui Counties were still generating significant income streams from their farming activities. Many rural households were supplementing their farming incomes with work in nearby towns. As a result, the rural households reported higher levels of income, financial security, income diversification, and food security than their rural-to-urban counterparts (Wilmsen 2018b).

These comparisons demonstrate the acute challenges faced by farmers who are resettled in urban areas. Land continues to provide an important safety net to displaced rural households, even as they increasingly engage with urban labour markets (Wilmsen 2016b; Zhan 2017). Land provides rural households with an important source of income, a means of growing their own food, a secure financial asset that can be leased out to earn income, and a base to which household members can return if jobs are lost or when they grow old. This safety net allows rural households to take advantage of off-farm work opportunities at a time of their own choosing. In contrast, rural-to-urban resettlers have no choice but to compete for work in urban labour markets, which often ascribe low value to their skills. These results also suggest that the Chinese government's focus on resettling more rural people to urban areas may drive the expansion of China's urban underclass (Wilmsen 2016b, 2018a).

Accounting for Non-respondents and Emigration

Our longitudinal research provided some valuable insights into the long-term impacts of resettlement at the Three Gorges Dam. A limitation of this research, however, was the 33% attrition rate. Some degree of attrition was to be expected, given the eight-year gap between surveys and the difficulty of tracking down households. Nonetheless, we wondered what had happened to the 178 households (of 521) who disappeared from our sample between 2003 and 2011 (surveys conducted in 2004 and 2012). We were also concerned that the most vulnerable households had been pushed out of the resettlement communities because they could not re-establish their livelihoods. If so, this would produce an overly optimistic assessment of long-term resettlement. The presence of such a dynamic would have implications for the existing resettlement literature more broadly. Academic studies as well as monitoring and evaluation programmes tend to focus on the initial years following resettlement. This narrow geographical and temporal focus reflects how resettlement is often framed as a two-phase, before-and-after process. As a result, little attention has been given to assessing the long-term viability of households' post-resettlement livelihoods and how this influences their propensity to emigrate in search of better opportunities elsewhere.

To address this gap, we returned once more to the field in 2016 to ascertain the whereabouts of the 161 missing households.[3] Using contact information collected in 2004, research assistants revisited the residences of the missing respondents. Data was collected from 88 (about 55%) of the missing households, that is, those that left the original resettlement site. Where research assistants could not contact the original respondents,

[3] Follow-up analysis revealed that of the 178 non-respondents in 2012, 17 households had passed away sometime between 2003 and 2012.

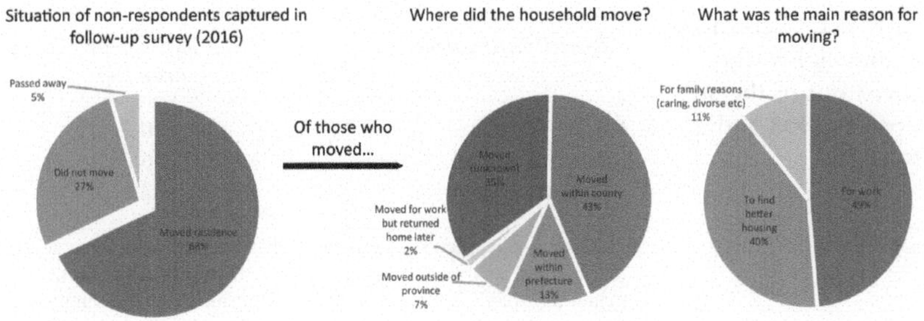

Fig. 3.1 Whereabouts of non-respondents in 2012 and their reasons for moving (N = 88)

they asked new occupants and neighbours about their whereabouts and reasons for moving.

Figure 3.1 summarises the key findings. A small proportion of the 88 households had passed away between 2012 and 2016 (5%). Another 27% of non-respondents were still living in their original houses but had not been home at the time of the 2012 survey (or had refused to participate). However, the majority of non-respondents (68%) had moved out of the resettlement area. Of those who moved, most had moved to a new residence within the same county (43%) or prefecture (13%). A small proportion of non-respondents moved outside the prefecture (7%) or moved away and subsequently returned (2%). The whereabouts of the remaining households (35%) could not be ascertained. The most important reasons given for moving out of the resettlement areas were to find work (49%) or better housing (40%). The remaining households (11%) moved for family-related reasons.

Importantly, the proportion of households who moved away from the resettlement area differed by resettlement category (Table 3.1). For example, 41% of rural-to-rural resettlers moved out of the resettlement area compared to 67% of rural-to-urban resettlers and 84% of urban-to-urban resettlers. These findings are consistent with those reported in the previous section; rural households were able to re-establish viable rural livelihoods when provided with adequate financial compensation and replacement land. In contrast, urban resettlers must find work in the local urban labour market and, failing this, move in search of work.

These data, combined with the findings of our previous research, suggest that the decision to leave the resettlement area was determined

Table 3.1 Percentage of households who moved out of the area between 2003 and 2012 (N = 88)

Resettlement category[a]	*Moved out of resettlement area (%)*
Rural-to-rural	40.7
Rural-to-urban	66.7
Urban-to-urban	83.7

[a]'Rural-to-rural' denotes households that relocated from a rural area to another rural area (usually uphill with replacement farmland); 'rural-to-urban' denotes households that moved from rural areas to urban areas (the local county town) and lost their farmland but gained urban ***hukou*** (household registration); 'urban-to-urban' denotes households that moved from the site of the old county town to the new county town

Source Author's creation

by an interaction between various push factors (unemployment, food insecurity, etc.) and pull factors (better housing and work opportunities elsewhere). Further, an extensive literature demonstrates household income is a factor that shapes a household's decision to emigrate (Zhu 2002; Yang et al. 2017). To explore this in more depth, we returned to our original dataset and analysed whether income and unemployment in 2003 was associated with non-response in 2012.

Our analysis provides evidence consistent with both push and pull dynamics. Table 3.2 demonstrates that the non-response rate in 2012 decreases as a function of household income in 2003 for rural-to-rural resettlers, suggesting some low-income rural households were pushed out

Table 3.2 Household incomes post-resettlement (2003) and non-response in 2012 (N = 88)

	Percentage of cohort that were non-respondents in 2012		
Household income per person (2003)	*Rural-to-rural (%)*	*Rural-to-urban (%)*	*Urban-to-urban (%)*
0–1000 RMB	39.4	22.2	41.2
1001–1750 RMB	25.0	23.9	33.3
1751–3300 RMB	21.4	15.4	36.1
3301–20,000 RMB	11.1	44.4	49.1
Sample average	28.7	22.3	40.3

Source Author's creation

of the resettlement area. The results for the urban resettlers, however, suggest the opposite relationship. Rural-to-urban and urban-to-urban households in the highest income category in 2003 were more likely than the cohort average to be non-respondents in 2012. Indeed, for the rural-to-urban cohort, households in the top income cohort in 2003 were nearly twice as likely as the cohort average to be non-respondents in 2012. This suggests that higher-income households with urban *hukou* were better placed to respond to employment and housing opportunities (i.e. pull factors) outside of the resettlement area than low-income urban households.

Unemployment

Local unemployment, combined with work opportunities elsewhere, is an important factor influencing a household's decision to emigrate (Yang et al. 2017). We therefore expect households with at least one unemployed adult in 2003 to be more likely to disappear from our sample than households in which all adult members were employed. Analysis of the data provides some support for this proposition, although it should be noted there were only 39 households in 2003 that reported an unemployed member (7.5% of the entire sample). Figure 3.2 summarises the jobs both prior to resettlement and, later, in 2011, held by those in the household who reported being unemployed in 2003. Prior to resettlement, the 39 unemployed household members were involved in a range of jobs, with farms and state-owned enterprises (SOEs) the two most important sources of employment. By 2011, however, 23% of households who had at least one unemployed member in 2003 reported that

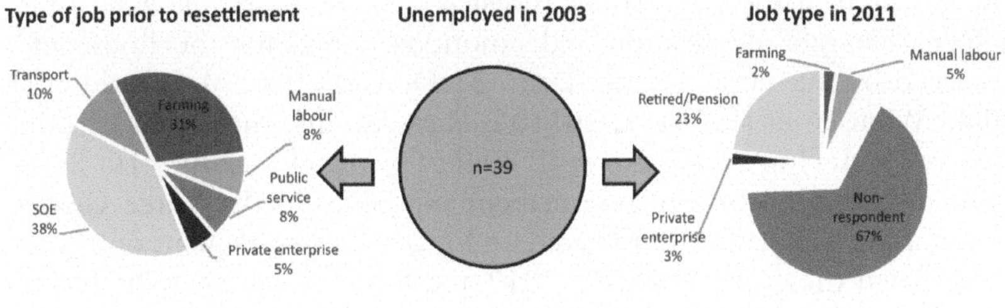

Fig. 3.2 Job types of those reporting being unemployed in 2003, pre-resettlement and in 2011

Table 3.3 Percentage of respondents and non-respondents in 2012 who reported at least one unemployed household member in 2003 (N = 39)

Percentage of respondents and non-respondents reporting that at least one member of their household was unemployed in 2003			
	Rural-to-urban resettlers (%)	*Rural-to-urban resettlers (%)*	*Urban-to-urban resettlers (%)*
Respondents in 2012 (reflecting on 2011)	1.7	4.6	6.0
Non-respondents in 2012	0.0	18.8	20.0

Source Author's creation

those members were retired or on welfare payments, highlighting the difficulties they had in finding new jobs after their resettlement. More importantly, 67% of households with at least one unemployed member in 2003 disappeared from our 2011 sample entirely.

Table 3.3 provides further evidence suggesting a link between unemployment in 2003 and non-response in 2012, but only for urban resettlers. This is consistent with the proposition that a small number of unemployed urban households were 'pushed' out of the resettlement area between 2003 and 2012 in search of work and, therefore, disappeared from our sample.

Discussion and Conclusion

Our research in the early 2000s revealed that the years following resettlement brought deprivation and emotional stress for the households displaced by the Three Gorges Dam (McDonald 2006; McDonald et al. 2008; Wilmsen et al. 2011a, 2011b). These results, supported by other early research (Jun 2000; Tan et al. 2005; Hwang et al. 2007; Tilt 2014) painted a bleak picture of resettlement outcomes in the Three Gorges. Recognising the paucity of research on long-term resettlement outcomes, we revisited these households in 2012 and found many households were doing surprisingly well. Whilst heartening, it is important to remember these improved outcomes were the result of the unique degree of attention given to promoting economic development in the Three Gorges region by the Chinese government.

One limitation of our longitudinal study was the attrition rate. This left us with some important questions. What happened to the missing households? And, more importantly, had some household left the resettlement areas altogether because they were unable to re-establish their livelihoods? To address these questions, we carried out a short follow-up survey in 2016 and discovered that the majority of missing households (68%) had moved out of the resettlement area. Of these, 49% had moved away for work-related reasons and another 40% had moved in search of better housing. We also found urban resettlers (i.e. those possessing urban *hukou*) were much more likely to have moved away from the area than rural households. We also analysed whether non-response rates in 2012 were associated with income and employment data collected in 2004. For rural-to-rural migrants, lower levels of income in 2003 were associated with higher non-response rates in 2012, suggesting push factors play an important role in out-migration of this cohort. For the rural-to-urban and urban-to-urban cohorts, there was evidence of push and pull factors. The association between unemployment in 2003 and non-response in 2012 suggests some urban resettlers were 'pushed' out of the resettlement communities in search of work. However, the income data found urban resettlers in the highest income cohort were more likely than the cohort average to be non-respondents in 2012, suggesting some wealthier households used their resources to take advantage of better housing and work opportunities elsewhere.

The quality of our data means we can only draw tentative conclusions about the magnitude and drivers of emigration from the resettlement area. Nonetheless, our findings point to important gaps in our own research and the resettlement literature more broadly. The data suggest that households reporting the greatest levels of disadvantage in 2003 were the most likely to disappear from our 2012 sample, suggesting that some had been pushed out of the resettlement area because they were unable to re-establish viable livelihoods. This means that our earlier findings may provide an overly rosy view of long-term resettlement outcomes in the Three Gorges. Furthermore, the existence of such a dynamic has been largely ignored in the existing resettlement literature. As such, the existing empirical record, which already paints a grim picture of resettlement outcomes around the world, may fail to account for the experiences of those who have been doubly displaced by resettlement projects. This lacuna reflects, in part, the fact that involuntary resettlement, as reflected in international standards of best practice, is treated as a before-and-after process imposed upon a well-defined and spatially circumscribed population. As a result, long-term resettlement outcomes are rarely measured—a gap our earlier work has sought to address—whilst the experiences of those who are forced out of resettlement communities are rarely captured.

More research is urgently needed on this neglected topic. This will require researchers to develop better strategies to track households over time and new theoretical frameworks for understanding households' motivations for leaving resettlement communities. Involuntary resettlement should be reframed as a local shock that ripples farther across time and space than is assumed in existing monitoring and evaluation practices. Greater attention should be given to how resettlement processes interact with normal life events (births, deaths, educational and employment opportunities, and the like) and other push and pull factors that influence a household's decision to 'stay put' in their resettlement communities. This, in turn, requires researchers and practitioners to engage more with the extensive literatures on inter- and intranational migration flows as well as other forms of forced displacement, such as refugee flows. With more than 200 million people projected to be displaced in the coming decades by large infrastructure projects (Cernea and Maldonado 2018), it is critical that we continue to deepen our theoretical and empirical understanding of involuntary resettlement and ensure that the mistakes of the past are not repeated.

Acknowledgements The authors acknowledge funding from the Australian Research Council, grant DE120101037.

References

Cernea, M. M. (1999). *The economics of involuntary resettlement: Questions and challenges*. Washington, DC: International Bank for Reconstruction and Development/The World Bank.

Cernea, M. M. (2016). The state and involuntary resettlement: Reflections on comparing legislation on development-displacement in China and India. In F. Padovani (Ed.), *Development-induced displacement in India and China: A comparative look at the burdens of growth*. Lanham: Lexington Books.

Cernea, M. M., & Maldonado, J. K. (2018). Introduction: Challenging the prevailing paradigm of displacement and resettlement. In J. K. Maldonado & M. M. Cernea (Eds.), *Challenging the prevailing paradigm of displacement and resettlement* (pp. 1–42). London: Routledge.

Chen, S. (2016). *From involuntary to voluntary: New-type urbanization approach to reservoir resettlement*. Society for Applied Anthropology Conference, Vancouver.

Department for International Development (DFID). (1999). *Sustainable livelihoods guidance sheets*. https://www.livelihoodscentre.org/documents/114097690/114438878/Sustainable+livelihoods+guidance+sheets.pdf/. Accessed 29 May 2020.

Duan, Y., & Steil, S. (2003). China Three Gorges Project resettlement: Policy, planning and implementation. *Journal of Refugee Studies, 16*(4), 422–443.

Duan, Y., & Wilmsen, B. (2012). Addressing the resettlement challenges at the Three Gorges Project. *International Journal of Environmental Studies, 69*(3), 461–474.

Duan, Y., Wilmsen, B., & Zhao, X. (2019). Urbanisation resettlement in China: Characteristics, risks and the revised land administration law. In S. Price & J. Singer (Eds.), *Country frameworks for development displacement and resettlement: Reducing risk, building resilience* (pp. 219–230). London: Routledge.

Han, X. (2018). *Money markets and hydropower: Chinese dam construction in Africa*. PhD dissertation, University of Melbourne.

Heggelund, G. (2004). *Environment and resettlement politics in China*. Hampshire: Ashgate.

Hensengerth, O. (2013). Chinese hydropower companies and environmental norms in countries of the Global South: The involvement of Sinohydro in Ghana's Bui Dam. *Environment, Development and Sustainability, 15*(2), 285–300.

Hwang, S., Xi, J., Cao, Y., Feng, X., & Qiao, X. (2007). Anticipation of migration and psychological stress and the Three Gorges Dam project, China. *Social Science and Medicine, 65*(5), 1012–1024.

International Hydropower Association (IHA). (2019). *China*. https://www.hydropower.org/country-profiles/china. Accessed 29 May 2020.

Jun, J. (2000). Environmental protests in rural China. In E. Perry & M. Selden (Eds.), *Chinese society: Change, conflict and resistence* (pp. 197–214). Oxon: Routledge.

McDonald, B. D. (2006). *From compensation to development: Involuntary resettlement in the People's Republic of China*. PhD dissertation, University of Melbourne.

McDonald, B., Webber, M., & Duan, Y. (2008). Involuntary resettlement as an opportunity for development: The case of urban resettlers of the Three Gorges Project, China. *Journal of Refugee Studies, 21*(1), 82–102.

Reddy, G., Smyth, E., & Steyn, M. (2015). *Land access and resettlement*. London: Routledge.

Rogers, S., & Wilmsen, B. (2020). Towards a critical geography of resettlement. *Progress in Human Geography, 44*(2), 256–275.

Scoones, I. (2009). Livelihood perpectives and rural development. *Journal of Peasant Studies, 36*(1), 171–196.

Scudder, T. (1997). Social impacts of large dam projects. In T. Dorcey, A. Steiner, M. Acreman, & B. Orlando (Eds.), *Large dams: Learning from the past, looking at the future* (pp. 41–68). Washington, DC: IUCN & World Bank Group.

Scudder, T. (2005). *The future of large dams: Dealing with the social, environmental and political costs*. London: Earthscan.

Smyth, E., Steyn, M., Esteves, A. M., Franks, D. M., & Vaz, K. (2015). Five 'big' issues for land access, resettlement and livelihood restoration practice: Findings of an international symposium. *Impact Assessment and Project Appraisal, 33*(3), 220–225.

Smyth, E., & Vanclay, F. (2017). The social framework for projects: A conceptual but practical model to assist in assessing, planning and managing the social impacts of projects. *Impact Assessment and Project Appraisal, 35*(1), 65–80.

State Council. (2006). *Regulation on land requisition compensation and resettlement of migrants for large and medium hydropower construction projects*. http://www.fao.org/faolex/results/details/en/?details=LEX-FAOC134774. Accessed 29 May 2020.

Tan, Y., Hugo, G., & Potter, L. (2005). Rural women, displacement and the Three Gorges Project. *Development and Change, 36*(4), 711–734.

Tang, S. (2016, March 11). China's new vision for the next five years. *China Today*, http://www.china.org.cn/china/NPC_CPPCC_2016/2016-03/11/content_38000088.htm. Accessed 29 May 2020.

Tilt, B. (2014). *Dams and development in China: The moral economy of water and power*. New York: Columbia University Press.

Urban, F., Siciliano, G., & Nordensvard, J. (2017). China's dam-builders: Their role in transboundary river management in South-East Asia. *International Journal of Water Resources Development, 34*(5), 747–770.

Vanclay, F. (2017). Project-induced displacement and resettlement: From impoverishment risks to an opportunity for development? *Impact Assessment and Project Appraisal, 35*(1), 3–21.

World Commision on Dams (WCD). (2000). *Dams and development: A new framework for decision-making*. London: Earthscan.

Wilmsen, B. (2016a). After the deluge: A longitudinal study of resettlement at the Three Gorges Dam, China. *World Development, 84*, 41–54.

Wilmsen, B. (2016b). Expanding capitalism in rural China through land acquisition and land reforms. *Journal of Contemporary China, 25*(101), 701–717.

Wilmsen, B. (2018a). Damming China's rivers to expand its cities: the urban livelihoods of rural people displaced by the Three Gorges Project. *Urban Geography, 39*, 345–366.

Wilmsen, B. (2018b). Is land-based resettlement still appropriate for rural people in China? A longitudinal study of displacement at the Three Gorges Dam. *Development and Change, 49*(1), 170–198.

Wilmsen, B., Adjartey, D., & Van Hulten, A. (2019). Challenging the risks-based model of involuntary resettlement using evidence from the Bui Dam, Ghana. *International Journal of Water Resources Development, 35*(4), 682–700.

Wilmsen, B., & van Hulten, A. (2017). Following resettled people over time: The value of longitudinal data collection for understanding the livelihood impacts of the Three Gorges Dam. *China. Impact Assessment and Project Appraisal, 35*(1), 94–105.

Wilmsen, B., van Hulten, A., Han, X., Adjartey, D. (2020). The environmental and social safeguard policies of the Belt and Road Initiative: The geopolitical implications. In M. Clarke, M. Sussex & N. Bisley (Eds.), *The Belt and Road Initiative and the future of regional order in the Indo-Pacific*. New York: Rowman and Littlefield.

Wilmsen, B., Webber, M., & Yuefang, D. (2011a). Development for whom? Rural to urban resettlement at the Three Gorges Dam. *China. Asian Studies Review, 35*(1), 21–42.

Wilmsen, B., Webber, M., & Duan, Y. (2011b). Involuntary rural resettlement: Resources, strategies, and outcomes at the Three Gorges Dam, China. *The Journal of Environment and Development, 20*(4), 355–380.

Wu, N. Z., & Liao, Q. L. (1999). The necessity to relocate the resettlers outside for the consideration of land bearing capacity—Take Yunyang County as a case study. *Resources and Environment in the Yangtze Basin, 8*(3), 20–22.

Xinhua. (2016). World's largest shiplift completes China's Three Gorges project. *China Daily*. https://www.chinadaily.com.cn/china/2016-09/19/content_26825380.htm. Accessed 29 May 2020.

Xu, X., Tan, Y., & Yang, G. (2013). Environmental impact assessments of the Three Gorges Project in China: Issues and interventions. *Earth-Science Reviews, 124*, 115–125.

Yang, Z., Cai, J., Qi, W., Liu, S., & Deng, Y. (2017). The influence of income, lifestyle, and green spaces on interregional migration: Policy implications for China. *Population, Space and Place, 23*(2), e1996.

Zhan, S. (2017). Hukou reform and land politics in China: Rise of a tripartite alliance. *The China Journal, 78,* 25–49.

Zhu, N. (2002). The impacts of income gaps on migration decisions in China. *China Economic Review, 13*(2), 213–230.

CHAPTER 4

Contestation Over Moral Economy: Distant Resettlement from the Three Gorges Area to the Pearl River Delta

Bettina Gransow

Introduction

In 2001, immediately upon their arrival in the Pearl River Delta (PRD), resettlers from the Three Gorges Dam project area (TGA) were visited by members of the local town government. On behalf of the provincial government, the representatives wanted to welcome the newcomers and present them with a gift of 100 Renminbi (RMB) per person. This was not part of compensation payments, but rather a friendly gesture by the provincial government and an expression of solidarity with the resettlers. Cheng (2006: 51) describes the scene as such:

> The local town government had set up an office in the resettlement village specifically for the village collective. On this day the meeting space was

B. Gransow (✉)
Freie Universität Berlin, Berlin, Germany
e-mail: bgransow@zedat.fu-berlin.de

J.-F. Rousseau and S. Habich-Sobiegalla (eds.), *The Political Economy of Hydropower in Southwest China and Beyond*, International Political Economy Series, https://doi.org/10.1007/978-3-030-59361-2_4

> full of resettlers. When the solidarity money was handed out, the resettlers became nervous. "Weren't we supposed to receive 8000 RMB of resettlement money?" "We were sent here as guests of General Secretary Jiang (Zemin) and Premier Zhu (Rongji), and this is how you want to treat us?"

The town leadership froze. They largely did not understand the Sichuan dialect and initially interpreted the emotional arousal as an expression of gratitude to the people of Guangdong. Gradually, the assembled leaders realised that the resettlers were cursing them; only when two of the resettlers corrected the matter did the others recognise the good intentions of the local government and the dark clouds passed (Cheng 2006). But from the beginning, the resettlers responded in unanticipated ways to the local host government's actions.

On their migration to South China, the resettlers—forty households comprising 134 people—had been accompanied by leading cadres of various ranks and regions, media coverage, and state propaganda. The group being displaced from the Three Gorges was made to feel special, as heroes who were sacrificing their small personal homes for their big national homeland (*she xiaojia, wei dajia*) (Cheng 2006). The discourse that rendered the displaced persons both heroes and victims was shared by the local host government, which was motivated not only by administrative obligation but also by compassion and commitment. The rationale behind the central government's long-distance resettlement scheme was that displaced people should be moved to downstream catchments of the Yangtze River benefitting from the Three Gorges Dam Project, located in economically developed provinces such as Guangdong (State Council 2001; Tan 2008).

Research on dam resettlement in China in general and on the TGA in particular has increased over the years (Li 2000; Heggelund 2004; Tan 2008; Price 2009; Wilmsen et al. 2011; Ying 2011, 2013; Peng and Feng 2015; Habich 2016; Padovani 2016; Wilmsen 2016; on distant relocation, see Xu and Shi 2001; Zhou et al. 2011; Liu and Wu 2016). However, there are very few longitudinal studies on the impact of displacement (though see Wilmsen 2016 on short-distance resettlement in the TGA and Wilmsen et al. in this collection). Of special note is the anthropological research by Cheng Yu on life in Baicun Village[1] during the early years of resettlement (Cheng 2006; Cheng and He 2006), which provided a rich baseline for preparing and interpreting the 26 in-depth interviews discussed in this chapter. These interviews, conducted in 2012, were part of a series of research projects on rural-to-urban migrant communities and the informal dynamics of megacity development in the PRD.[2]

To better understand the agency of the main actors and their different perceptions of the resettlement village, the present analysis uses the concept of moral economy. This concept was elaborated by E. P. Thompson (1971) in the context of food riots in the eighteenth-century English countryside, where peasants saw a traditional 'fair' price as more legitimate than the market price, inciting bread rebellions. James Scott (1976) later took up the concept with regard to Southeast Asian peasant societies, emphasising the role of shared moral values, group solidarity, and collective practices in avoiding subsistence crises for all villagers. This perspective was challenged by Samuel Popkin (1979), who presented the Asian peasant as an economically rational agent motivated by private interest. Following these classic studies, which have been labelled "the moral economy debate" (Little 1989: 29), the concept of moral economy was taken up by scholars of contentious politics in China. Kevin O'Brien and Li Lianjiang (2006) highlight ways in which moral economy overlaps and diverges with their influential concept of 'rightful resistance' in the context of rural contention in China. Recently, Bryan Tilt (2015) has used the concept of moral economy with a focus on water management decisions in China's Yunnan Province; he analyses the strategies of social interest groups and how these groups advance different moral visions of water resource management. Ying Xing (2013), who has done extensive empirical research on petitioning and collective actions in the TGA, discusses the idea of moral economy in relation to the Chinese concept of vigour (*qi*) in his study of contentious politics. Charles Watters (2013), building on E.P. Thompson's work on medieval notions of the 'deserving'

[1] Baicun is the pseudonym for the resettlement village which was our field site. This is the same pseudonym used by Cheng Yu in his earlier research (Cheng 2006). In the English publication by Cheng and He (2006), Baicun is translated as 'White Village'.

[2] The projects were embedded in a broader research programme entitled 'Megacities – Megachallenges: Informal dynamics of global change' and funded by the German Research Foundation.

poor in England, has broadened the understanding of moral economy as inherently linked to ideas of 'deserving' and 'undeserving'; his work is especially relevant to this chapter.

It is argued that perceptions and normative frameworks regarding the newly established resettlement village differed greatly among the local host government, the local host village, and the group of displaced people, with differences going beyond economic conflicts over compensation. After a short overview of the legal and policy background regarding distant resettlement from the TGA to Guangdong Province and the resettlement measures undertaken by the responsible town government, this chapter discusses the expectations and disillusions of the resettlers based on our 2012 interviews in Baicun. The analysis utilises the Impoverishment Risk and Reconstruction Model (IRRM) set forth by Michael Cernea (2000)[3] as well as earlier findings by Cheng (2006).

Legal and Policy Background of Distant Resettlement from the Three Gorges Dam Area

TGA resettlement was guided by the policy of 'developmental resettlement' (*kaifa xing yimin*). Proposed by the Chinese government in the mid-1980s, this policy had two objectives: (1) solving ongoing problems such as large-scale impoverishment faced by reservoir resettlers from earlier dam projects, and (2) preparing for TGA resettlement (Tan 2008). Article 3 of the 'Regulations on Resettlement in the Yangtze Three Gorges Project', promulgated by the State Council in 1993, and revised in 2001 and 2011, stated that to create conditions for economic and social development in the reservoir area, the Three Gorges project would implement a developmental resettlement guideline, coordinate use of resettlement funds, show reasonable development of resources, protect the environment, make proper arrangements for resettlers, and enable resettlers' production and lives to achieve or exceed their original standards (State Council 2001). This broad and rather vague formulation gave rise to questions and controversial interpretations of the concept of developmental resettlement. What exactly does it mean to say that resettlers' production and lives should achieve or exceed the original standards?

[3] The IRRM provides a framework for risk assessment in resettlement processes, particularly focusing on the risk of impoverishment, which has been recognised as the most widespread effect of involuntary resettlement.

Regarding poor households—should they be left as poor as they were before?[4]

In 2000, an important policy shift led to rural households being resettled in more distant sites rather than higher-elevation locations within the immediate reservoir area. This adjustment largely shifted the responsibility for restoring and developing resettlers' livelihoods to the distant host governments. According to official statistics, 1.27 million people were displaced from the TGA from 1993 to 2009, of whom 196,000 were resettled to distant locations. The latter figure can be divided into 35,000 self-organised distant resettlers who moved to family or friends throughout the country, and 161,000 government-organised distant resettlers who were sent to 11 provinces or province-level municipalities (Zhou et al. 2011).

Also in 2000, the government of Guangdong produced the No. 123 document that specified the distribution of 7006 distant resettlers among the four cities of Foshan, Huizhou, Zhaoqing, and Jiangmen.[5] Distribution was to be concentrated in municipalities with township (*xiang*) and town (*zhen*) administrative levels and scattered among individual villages. The distributing cities (*shi*) were to select six or seven towns under their administration, and each town in turn was to select three to five suitable sites which were each to take 10 households or approximately 40 people (Guangdong 2000). In the case discussed below, the town government selected Chen Village[6] on behalf of the district government as the host location for the new Baicun Resettlement Village—but here the top-down chain ran into problems.

Based on the directives from the provincial government, the rural resettlers in Guangdong were to be accommodated in accordance with the principle of 'moving out, steady housing, gradually getting rich'

[4] Interestingly, when the concept of developmental resettlement was taking shape in China in the 1980s and 1990s, the World Bank, a pioneer in drafting involuntary resettlement policies (and which refused to help fund the Three Gorges Dam because of the project's high risks), watered down its progressive definition of development-oriented resettlement provided in 1990 in the Operational Directive 4.30. Its 2001 version ended up allowing borrowers the option of merely restoring incomes, which in turn could result in impoverishing the majority of resettlers (Tan 2008).

[5] But in fact, from 2000 to 2006, 9007 rural residents from the TGA area were displaced to Guangdong (Tan 2008).

[6] Chen Village is not the real name of the village; as an administrative village (*guanli cun*), Chen Village consists of several natural villages (*ziran cun*).

(*ban dechu, wen dezhu, zhubu neng zhifu*). Resettler livelihoods were to be centred on land and agriculture, and local governments were to ensure that the newcomers received contracted arable land (*chengbao de gengdi*), private plots (*ziliudi*), and homesteads (*zhai jidi*) of a quantity and quality not below the average level of local farmers. Based on the strategy of development-oriented resettlement, the newcomers were to contribute to farming, forestry, animal husbandry, fishing, and associated rural industry in accordance with local conditions, and were encouraged to engage in the production and service sectors in order to increase their incomes on a step-by-step basis. Counties (*xian*) and cities (*shi*) considered suitable for resettlers possessed good natural conditions, sufficient land, strong leadership, and the ability to provide sufficient infrastructure (water, power, roads) as well as schools, hospitals, and other public facilities. The resettlement locations were to be in the vicinity of market towns and were also to have convenient working, living, and transportation conditions for the resettlers. Accommodation was to be based on the principle of "three times one" (*san ge yi*), i.e. each household was to have one piece of arable land, one house, and one person joining the outside labour force (Guangdong 2000).

Resettlement Measures Undertaken by the Local Town Government

In accordance with the requirements of the Guangdong government, the town government made an agreement with Baicun Resettlement Village that included the following ten points (Cheng 2006: 57–58):

1. The resettlers will receive 0.6 *mu* arable land and 30 m^2 homestead land per capita.
2. House construction will be calculated at a cost of 480 RMB per m^2, of which the resettlers have to bear 250 RMB. According to the national Three Gorges resettlement plan, each resettler will get 20 m^2 of housing, which means a subsidy of 4600 RMB.
3. The resettlement village will be provided with infrastructure (electricity, water, roads, and postal service).
4. The government will bear the fees for the household registration booklet, the house title deed (*fangchan zheng*), and the land document (*tudi zheng*).

5. Resettlers will receive guidance and training for setting up production.
6. The government will provide 3875 RMB to each resettler during the transitional period, consisting of a monthly living allowance (100 RMB per capita for 18 months) and a lump sum for purchasing production materials. Households in economic difficulty will receive a subsidy based on procedures in the related document from the State Council.
7. Resettlers who meet the conditions will receive assistance in finding factory work.
8. After registration, school fees will be reduced by 50% for at least 3 of the 9 years of compulsory schooling, as well as for upper-middle school students.
9. After becoming legal members of the village, resettlers will have the same rights and obligations as local villagers.
10. Based on national and Guangdong regulations for resettlers and starting from the year of resettlement, income from non-wasteland agriculture and from special local agricultural products will be tax-free for 3 years.

The compensation and accommodation costs were shared by the county government in the Chongqing reservoir area that sent the Baicun resettlers and the district government of the city that received them. The compensation standards for the Three Gorges resettlers were higher than those for any previous reservoir resettlement project in China. However, as we know from resettlement research, losses span much more than income alone. In his analysis of the Baicun villagers' initial years after relocation, Cheng (2006) used Cernea's IRRM (2000) to evaluate safeguarding measures taken by the local government. The IRRM has a dual emphasis on risks to be prevented or mitigated and on reconstruction measures to be implemented. According to this model, involuntary resettlement may cause eight kinds of risks, each of which have economic and social dimensions. Following Cernea's model, Table 4.1 summarises the measures taken by the receiving town government and Cheng's comments on whether the problems identified in the model matched the situation in Baicun Village and could be solved accordingly.

Overall, the town government did not have an easy task. On the one hand, they wanted to fulfil the requirements of higher-level governments (district, city, province levels) and provide suitable resettlement measures

Table 4.1 Assessment of resettlement measures by the town government in comparison with the IRRM

Social risk IRRM	*Reconstruction measure IRRM*	*Reconstruction measure by local government*	*Comment (Cheng 2006)*
1. Landlessness	Land-based resettlement	Land-for-land compensation	Peri-urban resettlers had worked in the tourism industry, not in the fields
2. Joblessness	Reemployment	Factory jobs with low wages	19 people left Baicun Village before 2003 and returned to TGA
3. Homelessness	House construction	New houses	Feeling of 'placelessness'; rented rooms to migrant workers
4. Marginalisation	Social inclusion	Initial 'welcome culture' from local host government	Rejection by the formal host village
5. Food insecurity	Adequate nutrition	Various compensation measures	Food insecurity was not a problem
6. Increased morbidity and mortality	Increased health care	Health infrastructure	Stomach problems, tuberculosis; lack of public health; traffic accidents
7. Loss of access to common property resources	Restore community assets and services	–	Loss of Three Gorges landscape and work in local tourism industry
8. Social disarticulation	Network-building and community rebuilding	Encouraged the election of a village head	Lack of cohesion among resettlers from different places

Sources Cernea (2000) and Cheng (2006: 175–182); Author's creation to the Baicun resettlers. On the other hand, they had to deal with the stubborn Chen Village committee, who—against the intentions of the town government—refused to host the Baicun resettlers. As a result, officials had to deal directly with the resettlement village on numerous issues. The town government tried hard to address more or less all the risks listed in Cernea's IRRM, and it was expected that the Baicun resettlement villagers would be quite satisfied with their new situation—after all, the PRD was a fast-developing region that seemed to fit perfectly with 'development-oriented resettlement' by providing considerable economic opportunities to the people displaced from the TGA.

Displacement from and to Peri-Urban Spaces: Expectation and Disillusion

The group of Baicun resettlers came from urban and peri-urban areas in mountainous regions on the banks of the Yangtze. Although their household registrations listed them as rural, most of the Baicun resettlers had lived right next to the county seat and the tourism industry was their main source of income. The families generally set up stands in front of their homes or opened small shops where they sold local specialties or souvenirs. Some of those who had attained higher education levels worked as guides and could earn several thousand RMB a month. Hardly anyone worked in the fields. Some grew a few vegetables and sold them in the city, but their land was scarce, steep, hard to cultivate, and located in an area of otherwise high population density. Regardless of their actual occupation, though, Baicun resettlers were formally considered peasants (Cheng 2006).[7]

[7] Two types of rural settings were identified in the TGA: purely rural agrarian areas and mixed peri-urban agricultural zones. There were significant differences in production, income sources, and land-use types between these two zones. Nonetheless, people living in non-agrarian peri-urban zones of a city or town were compensated and resettled like other rural residents engaged in agricultural production (Tan and Wang 2003). The criterion was the rural household registration (*nongcun hukou*) of these families rather than their actual occupations, sources of income, or lifestyles.

Remembering Displacement and Arrival

Ms. Wang,[8] 47, a villager from Baicun resettlement village, told us about her former situation in the Three Gorges area: "I used to live in the county seat. I had a rural *hukou*, no work, and no land. We were called peasants, although we weren't peasants. We were city dwellers, although we weren't called city dwellers". Mr. Zhou, 49, another Baicun villager, now working as security coordinator, described the situation in similar terms: "Our house was in the city centre, although we had a rural *hukou*. We were [at the] banks of the Yangtze, very close. Urban people were removed to the back [to new development areas behind the city] whereas the rural inhabitants were resettled far away. I also didn't want to leave then".

The first two years after arriving in Guangdong were described in retrospect by the resettlers as the most difficult period, in which they experienced multiple shocks. As Mr. Zhou explained: "The first step was very hard. You think you're being resettled to a metropolis, and then you see that there's a lot of garbage lying around and you yourself have to take care of removing it". The new location was considered remote, with the county seat 10 km away. Ms. Lu, 39, a Baicun housewife, remembered: "At the beginning everything outside was full of fish ponds, there wasn't a single factory. Now things are better". The Baicun resettlers saw joblessness as a much greater problem than landlessness. Because most had worked in the service sector, including the tourism industry, they generally lost their businesses and clients. The new environment devalued their labour qualifications and experience. Land-for-land resettlement for people from the peri-urban areas in Chongqing meant moving to a desolate, newly-founded resettlement village surrounded by fish ponds, which from their perspective was more of an urban-to-rural resettlement than the anticipated move to an economically more developed region. For this and similar reasons, 19 people left Baicun Village before 2003 and went back to the TGA.

The Use and Value of Compensation Houses

Most of the resettler families received one- or two-storey houses in compensation for the apartments or houses they had lost. The objective

[8] Pseudonyms are used for interviewees.

was to prevent them from becoming homeless as a result of resettlement, and to give them new homes. As it turned out, the resettlers were interested not only in the practical value of living in the houses themselves. They also saw them as possible sources of additional income, especially by renting out rooms or apartments. In contrast to the bungalows in other resettlement villages in the PRD, the houses had such good foundations that one or two additional storeys could easily be added to them. As Ms. Guo, 40, who worked in a toy factory after arrival in the PRD, highlighted: "Generally everyone rents out, and lives above with their families. They rent to migrant workers, who come from everywhere. It costs 120–150 RMB a month to rent a room". That was not the case at the beginning. The resettler families then lived on the ground floor, as that made it easier to engage in conversation with passers-by. Mr. Zhang, 43, regretted that he could not realise his initial plan of just selling his compensation house after arrival, complaining: "I don't go out very often. I used to live below, and everything was better, everyone came to the door. Now everyone lives on the upper floors, so you can't just stop by". But rentals are essentially never the sole source of income. Instead, they are seen as additional financial security in the event of unemployment, or for old age care, or to pay school fees for children.

Not every family in Baicun could come up with the money needed to enlarge their houses, and some found that they could not use their house as collateral for securing a bank loan to buy a motorcycle or car. Ms. Li, 32, another Baicun housewife, was outraged:

> When we wanted to take out a loan, they said we had no fixed assets, they totally ignored us. We turned to the government, but in vain, they said they couldn't do anything about it. Local people may get 180,000 (RMB) without any mortgage. I said we also have a local household registration, then they said, you are different – one sentence and they sent us out. It's not that we don't integrate into this society, it's the local government, ah, and institutions which simply don't take us as one of their members. Real integration is not an easy thing, the local people here are too xenophobic. This becomes most obvious when they don't treat us as people from here.

More than ten years after arriving in Guangdong Province, the inhabitants of Baicun Resettlement Village were not associating their compensation houses with the idea of 'belonging' or 'home' in the PRD. The houses were perceived and used as a place to live and literally as a resource to build on. The exchange value of the houses seemed to be at least as important as their use value to the resettlers, who had agreed to move to a distant location as a sacrifice to the nation and who had expected a considerable increase in income as compensation for all their losses. Their moral economy differed considerably from that of the government, which considered its job done after it provided solid housing that would keep the resettlers from homelessness.

The Use and Value of Compensation Land

A similar difference between the moral economy of the government and that of the resettlers can be seen with regard to land compensation. The local town government wanted to implement the resettlement policy of the district, city, and provincial governments and fulfil its contractual obligations to the Baicun villagers. This was not an easy task for them—arable land was a scarce and precious resource, and the idea of passing their responsibility down to the administrative village, Chen Village, was not met with any kind of appreciation in this locale. The only concession the town government was able to negotiate with Chen Village (without taking compulsory measures) was the allocation of 0.2 *mu* arable land plus 0.4 *mu* fish ponds per capita for the 134 residents of Baicun (Cheng 2006: 175).

Because most of the Baicun resettlers had no background in agricultural production they did not really care about the compensation land, at least not to begin with. Ms. Wu, who worked in a nearby factory and shared with her husband a room rented from her uncle, a Baicun villager, explained: "Many people in the village don't know how to farm the fields. Because many people from the county seat didn't work in the fields, in Chongqing they did not farm. When they decided to come here, it was definitely not to work in the fields. Moreover, the fields are quite far away". Besides getting much poorer land than that of the Chen villagers, the Baicun resettlers were also wondering why the compensation land they received only totalled 0.6 *mu* per capita. Even though this was more than what they had possessed before resettlement, Chen villagers themselves owned 2.6 *mu* of arable land per capita. For Baicun resettlers who perceived the quantity of compensation land not least as a status symbol in their new environment, this difference in land ownership came with the perception of being assigned a lower social status than the Chen villagers (Cheng 2006: 175).

Land usage rights were subsequently sold to a coal plant to use as a stockyard. Ms. Lu pondered this, saying: "The land has been sold, but the problems remain. I think we sold the land too cheaply. We sold one *mu* for 36,000 RMB... The coal plant back there leased it". She also wondered how to profit from the homestead land she received: "Homestead land is 10 m^2 per person. For our family this would mean 50 m^2. We could build housing and rent it out. When we get older and don't work anymore, then we'd still have the rental income—but we don't have the money now to build". Here, too, more than ten years after the resettlers had arrived, the question of the homestead land and its use had still not been settled with the town government. Similar to the housing situation, resettlers' compensation land turned into a resource with an exchange value—that is, a resource the resettlers could use to generate income.

Chen Village: The Host(ile) Village

Though Baicun is formally situated within the political boundaries of Chen Administrative Village, it lies at a somewhat higher elevation and is clearly separated and visually distinct from the rest of Chen Village. Its walls reinforce the impression of a separate, closed place. Chen Village refused to include Baicun as one of its village working groups on the grounds that the collective economy of Chen Village was very strong, as opposed to that of Baicun. At the end of each year, virtually all of the affiliated natural villages would contribute surpluses to the administrative village, which then distributed a jointly earned dividend (*fenhong*). Because the resettlement village had no collective economic resources and could not contribute anything to the collective contract, it would only receive dividends. Chen Village dwellers therefore decided to vote against accepting Baicun. Under pressure from the town mayor (*zhenzhang*), they ultimately agreed to sign the documents required by upper-level governments to accept Baicun at least on formal terms (Cheng 2006).

Chen villagers characterised the Three Gorges resettlers as people spoiling for a fight, and wished to avoid them (Cheng 2006). They described Baicun as dirty and the newly built houses as making a messy impression (Cheng 2006). Chen Village, in contrast, was seen as a peaceful, clean, and quiet place. With their critical and demeaning view of Baicun villagers, Chen villagers also underscored their self-image as an economically and socially intact community to justify their disdain. Baicun villagers felt discriminated against by this attitude. In the eyes of Mr.

Sun, 40, a Baicun villager working in a ceramics factory, the relationship between the two villages underwent some changes over the long term. He described his impressions as such: "It was unpleasant at first, when people were condescending toward you; it was as if you were somehow being discriminated against or even shoved to the side. Now the attitude is that you don't owe me anything and I don't owe you anything". This statement may be interpreted as the emerging urban indifference of the metropolitan person, in the sense of Simmel (1998 [1903]).[9]

Baicun: From Resettlement Village to Diaspora Village

The resettlers who suddenly became Baicun villagers in mid-August 2001 had lived in four different villages before relocation, and therefore had only a limited sense of mutual trust and security (Cheng 2006; Cheng and He 2006). Ms. Lu remembered:

> When we arrived, I didn't know the other people, just three families that came from the nearby area; we were acquainted with them, like neighbours. The other resettlers all came from the suburbs of the county seat, and our area was more rural. We only got to know the other resettlers after we got here. Now the relations in the entire village are very good, like members of a family.

Mr. Zhu, 52, a Baicun villager involved in the management of construction projects, described the composition of the village: "Overall, one can say there are three systems: the biggest would be those who belong to the Tao family. A lot of people have the Wei family name, then their relatives on both sides, those who are connected to them, close relations. The others are like us, individual households, without relatives or with only a few". Lack of cohesion and social disarticulation among Baicun villagers manifested itself in frequent, noisy quarrels that arose from trivial issues and quickly turned into conflicts.

Baicun Village is neither a natural village nor an administrative village; it is a special resettlement village (*yimin cun*) under the administration of the local town government. It is therefore not represented by a village

[9] In his classic 1903 sociological study, Georg Simmel (1998 [1903]: 121) characterised social relations in the city as "indifferent to everything that is individual".

committee leader (*cunweihui zhuren*) but rather by an elected representative called the village head (*cunzhang*) (Cheng 2006). In the beginning, the town government displayed an accommodating attitude towards the newly elected 24-year-old female village head. The town sought to let Baicun assume more organisational work in hopes that this would increase social cohesion and relieve the town government from needing to address Baicun villagers' concerns on each and every issue.

At first, Baicun villagers only swept in front of their own house doors (Cheng 2006). Garbage heaps were growing, while the electric street lights did not work and no one seemed to care. The town government thus urged the village head to channel some of the revenues that the Baicun villagers had received from a fish pond contract with Chen Village into a joint fund to care for public spaces. The town mayor highlighted the generosity of Chen Village in transferring the fish ponds' entire 2001 revenue, even though Baicun residents had only arrived in mid-August. However, under pressure to prove to the villagers that she was not lining her pockets but rather standing up for their wishes, the village head resisted, insisting that the money be distributed to individual Baicun households. Though she and the villagers prevailed, after this snub of the town mayor, the "welcome culture" phase came to a definitive end and a phase of "cold war" began (Cheng 2006: 103–104; see Cheng and He 2006 for more detail about the election of the first village head). Soon after, the village head resigned from her post and the local government established an administrative office for the Three Gorges resettlers. However, the office lacked special staff or designated office space and therefore only made the local government more opaque and difficult to approach. Cadres from the local government avoided visiting Baicun Village and only came for special occasions such as the tenth anniversary celebrations in 2011. This was about when the village got a new wall—for both protection and surveillance. Ms. Gao, 51, was sceptical: "[The wall is] for security. Each village has to build one, to install the cameras. The government provides 30,000 RMB, the rest should be paid by us. Our village has no money, they said we should build [a wall], so we are building one, [but] we lack the money…." Attempts by the local government to foster a common sense of responsibility among the resettlers were met with little response. Shared projects and investments did not move beyond the initial consideration phase for years. Over the long term, the resettlement space that the government had provided did not lead to increased social cohesion among villagers, nor did it allow them to integrate into local society.

Overall, social relations and networks extending beyond the resettlement village were found to be rare. With the exception of schoolchildren, Baicun residents had few friendships with people from Guangdong. There was also hardly any contact with other resettlement villages in the PRD. Even though overall living standards improved in Baicun over the eleven years—for instance, ten families owned a car in 2012—and tensions with Chen Village decreased, isolation and an uncertain status remained. Based on their own self-image, Baicun villagers saw their identities not as local immigrants but rather as a diaspora of Three Gorges resettlers.

Conclusion

The moral economy of the Three Gorges resettlers who became villagers in the PRD was shaped by a mixture of expectations. Resettlers aimed to be recognised as heroic figures that facilitated the prestigious national dam project and considered themselves deserving of fair compensation for their losses and a fair share of the development opportunities available in the local host society. With shattered hopes of recognition and limited job opportunities, the resettlers tried to generate additional income from the houses and compensation land they received. However, with no social networks in Baicun, they had to resort to the local government, which they approached individually on an ad hoc basis until they got on local cadres' nerves. As they had moved for the national good, they felt that they deserved material as well as social recognition, as well as the chance to enhance their livelihood opportunities as full-fledged Guangdong citizens.

The local government perceived the resettlement space provided to the Three Gorges resettlers as a place to stay and eventually feel at home. They expected that resettlers to self-organise and handle community issues and public spaces by themselves. From a moral economy perspective, the local town government had a different understanding of 'deserving' versus 'undeserving'. Local officials saw it as their duty to provide solid houses, compensation land, jobs, and basic material and social infrastructure for the newcomers. They expected a grateful and quiet resettlement village to emerge in return. In the beginning, authorities showed compassion with the 'deserving' resettlers. But this attitude changed when the resettlers did not respond as expected, and when resettlers turned into a disorganised group of noisy and demanding villagers who insulted them. Instead of searching for the root problems of such behaviour (such as discrepancies between the resettlers' new 'rural' status and the urban lifestyles they were accustomed to, as well as the devaluation of their former sources of income) the local government retreated from the resettlement village. Authorities considered their obligations towards Baicun Village fulfilled, but had to keep addressing administrative issues stemming from upper-level governments as well as with those from Chen Village, whose inhabitants' negative attitude towards the resettlement village did not evolve much over the years.

Contestation over the resettlement village arose from the differences in how local governments, the Baicun villagers, and Chen Village conceptualised this space. The local government wanted to see Baicun Village as an internally integrated community where people would develop some sense of belonging; this would in turn bring inhabitants to acknowledge officials for having carried out their task successfully. Chen Village saw the resettlers as 'others' infringing upon their resources; they wanted the resettlement space to be limited, separated from Chen Village, and as invisible as possible. The Baicun villagers themselves used the resettlement space as a resource; they generated additional income by renting out rooms and some hoped to sell the houses and eventually go back 'up' to Chongqing. Their emerging identity became shaped more by their status as Three Gorges resettlers than by their new formal status as Guangdong citizens.

References

Cernea, M. (2000). Risks, safeguards and reconstruction: A model for population displacement and resettlement. In M. Cernea & C. McDowell (Eds.), *Risks and reconstruction: Experiences of resettlers and refugees* (pp. 11–55). Washington: World Bank.

Cheng, Y. (2006). *Life in White village: An anthropological study of the adaptability of three gorges resettlers in Guangdong Province*. Beijing: Minzu Press.

Cheng, Y., & He, X. (2006). Power games and migrant adaptability in migration villages: The case of White Village, a settlement in Guangdong Province for migrants displaced by the Three Gorges Dam project. *Chinese Sociology and Anthropology, 38*(3), 71–89.

Guangdong. (2000). *Document number 123 of the Office of the People's Government of Guangdong Province*. http://www.gd.gov.cn/gkmlpt/content/0/137/post_137063.html#7. Accessed 3 June 2020.

Habich, S. (2016). *Dams, migration and authoritarianism in China: The local state in Yunnan*. London: Routledge.

Heggelund, G. (2004). *Environment and resettlement politics in China: The three Gorges project*. Aldershot: Ashgate.

Li, H. (2000). *Population displacement and resettlement in the Three Gorges reservoir area of the Yangtze River, central China*. PhD dissertation. University of Leeds.

Little, D. (1989). *Understanding peasant China: Case studies in the philosophy of social science*. New Haven: Yale University Press.

Liu, J., & Wu, D. (2016). Analysis of social adaptability in the Three Gorges distant resettlement. *China Water Transport, 16*(10), 60–62.

O'Brien, K. J., & Li, L. (2006). *Rightful resistance in rural China*. Cambridge: Cambridge University Press.

Padovani, F. (Ed.). (2016). *Development-induced displacement in India and China: A comparative look at the burdens of growth*. Lanham: Lexington Books.

Peng, H., & Feng, G. (2015). *Study on the social adaptability of Three Gorges resettlers*. Wuhan: Wuhan University Press.

Popkin, S. L. (1979). *The rational peasant: The political economy of rural society in Vietnam*. Berkeley: University of California Press.

Price, S. (2009). Prologue: Victims or partners? The social perspective in development induced displacement and resettlement. *Asia Pacific Journal of Anthropology, 10*(4), 266–282.

Scott, J. C. (1976). *The moral economy of the peasant*. New Haven: Yale University Press.

Simmel, G. (1998 [1903]). Die Großstadt und das Geistesleben. In G. Simmel (Ed.), *Soziologische Ästhetik* (pp. 119–133). Frankfurt: Philo Verlagsgesellschaft.

State Council. (2001). *Regulations on resettlement in the Yangtze Three Gorges project*. Beijing: Chinese Legal Press.

Tan, Y. (2008). *Resettlement in the Three Gorges project*. Hong Kong: Hong Kong University Press.

Tan, Y., & Wang, Y. (2003). Rural resettlement and land compensation in flooded areas: The case of the Three Gorges project. *China. Asia Pacific Viewpoint, 22*(1), 35–50.

Thompson, E. P. (1971). The moral economy of the English crowd in the eighteenth century. *Past and Present, 50*(1), 76–136.

Tilt, B. (2015). *Dams and development in China: The moral economy of water and power*. New York: Columbia University Press.

Watters, C. (2013). Forced migrants: From the politics of displacement to a moral economy of reception. In S. J. Gold & S. J. Nawyn (Eds.), *The Routledge international handbook of migration studies* (pp. 99–106). New York: Routledge.

Wilmsen, B. (2016). After the deluge: A longitudinal study of resettlement at the Three Gorges Dam. *World Development, 84*, 41–54.

Wilmsen, B., Webber, M., & Yuefang, D. (2011). Development for whom? Rural to urban resettlement at the Three Gorges Dam, China. *Asian Studies Review, 35*(1), 21–42.

Xu, J., & Shi, G. (2001). Three Gorges long-distance resettlement and economic integration in coastal resettlement areas. *Modern Economic Research, 11*, 3–6.

Ying, X. (2011). *"Vigour" and contested politics: A study of the problem of stability in contemporary rural Chinese society*. Beijing: Social Sciences Academic Press.

Ying, X. (2013). *A study of the stability of contemporary rural Chinese society*. Berlin: Springer Verlag.

Zhou, Y., Zhou, Y., Hu, H., & Xiang, C. (2011). The regional differentiation of the Three Gorges rural emigrants adaptability and the revelation. *Economic Geography, 31*(8), 1351–1357.

CHAPTER 5

Population Resettlement for Hydropower Development in the Lancang River Basin: An Evolving Policy Framework and Its Implications for Local People

Bryan Tilt and Zhuo Chen

Introduction

With more than 25,000 large dams, China leads the world in hydropower production. As hydropower meets a larger share of China's escalating energy demands, it may also help to reduce the nation's reliance on fossil fuels; government agencies and private entities alike are pursuing alternative energy-development plans involving not just hydropower, but also wind, solar, wave, and biogas. The development of a so-called low-carbon economy (*ditan jingji*) is welcome news in a country where hundreds of thousands of people die each year from ailments linked to air pollution from fossil-fuel combustion, where pollution-related economic losses

B. Tilt (✉) · Z. Chen
Oregon State University, Corvallis, OR, USA
e-mail: Bryan.Tilt@oregonstate.edu

J.-F. Rousseau and S. Habich-Sobiegalla (eds.), *The Political Economy of Hydropower in Southwest China and Beyond*, International Political Economy Series, https://doi.org/10.1007/978-3-030-59361-2_5

cut into the nation's gross domestic product, and where aggregate levels of greenhouse gas emissions are the highest in the world (GDB MAPS Working Group 2016).

However, China's push for hydropower development has attracted national and international scrutiny because of a range of ecological and social problems that seem endemic to this form of energy production: fragmentation of riparian ecosystems, species extirpation, water-quality degradation, and the displacement of individuals and communities unfortunate enough to live in the path of large-scale infrastructure development. When dams are built and reservoirs fill behind them, they displace the human beings who live there, flooding farmland, inundating homes, and causing a range of social and economic problems for communities. The effects can last for generations as people cope with the consequences for their families' income, their way of life, and their sense of place and community (Scudder 2005). China's hydropower sector thus highlights the tensions and trade-offs between economic development, energy production, and social welfare.

In this chapter, we examine hydropower development on China's Upper Mekong River (hereafter referred to by its Chinese name, Lancang), with a particular focus on the consequences for displaced communities, many of whom belong to minority ethnic groups or other historically marginalised populations. Our analysis draws on a review of policy documents from central and provincial government agencies and hydropower corporations, along with surveys and interviews conducted in 2010 by American and Chinese researchers with rural households in more than 30 villages across Fengqing, Yun, and Lancang Counties in Yunnan Province (the site of four dams built along the mainstream of the Lancang River between 1996 and 2014). We supplement these earlier findings with a review of recently published academic articles and policy documents on the Miaowei Hydropower Project, which was completed in 2017. Resettlement for the Miaowei Project was conducted under a new resettlement policy called 'Opinion on Compensation and Subsidies for Resettled People on the Middle Reaches of the Jinsha River for Hydropower Development'. Drafted in 2007 and subsequently adopted throughout China as the new standard for hydropower resettlement practice, this policy has become colloquially known as the '16118 policy', a numerical abbreviation for the new practices, procedures, and compensation standards, which we describe in detail below. This allows us to analyse in some detail how newer hydropower projects on the Lancang

River differ from their predecessors in terms of population resettlement and its social impacts, although, as we describe below, direct comparisons are sometimes difficult because of the wide variation across these policy regimes.

Hydropower Development on the Lancang River

China's hydropower expansion is driven by remarkable growth in energy demand throughout the domestic market and by central government directives promoting new sources of energy in order to sustain economic growth, ensure energy security, and promote environmental protection. Although the pace of construction on new hydropower facilities has slowed, primarily because the most technically and economically feasible sites have already been developed, the social consequences of displacement and resettlement continue to be felt by individuals and communities. The southwestern province of Yunnan has been designated as one of thirteen key 'hydropower bases' for the nation, and the Lancang River, which rises on the Qinghai-Tibet Plateau at an elevation of 5224 m, figures prominently in the region's hydropower development plans. The Lancang River flows from north to south through Yunnan Province for 2200 km. In Yunnan alone, the Lancang watershed is home to more than 20 of China's officially recognised 'minority nationalities', ethnic groups who often endure economic and cultural marginalisation and whose standard of living lags behind much of the rest of China (UNDP 2008). The middle and lower segments of the Mekong (as it is known in Southeast Asia), along with an extensive network of tributaries, support fishing and floodplain agriculture for tens of millions of people in five downstream riparian countries including Myanmar, Laos, Thailand, Cambodia, and Vietnam (Grumbine et al. 2012).

Hydropower development plans for the Lancang River have evolved considerably in recent years. The lower section in Yunnan Province includes a lower cascade of seven extant dams: Manwan (132 m tall, 1.5 GW installed capacity, completed 1996); Dachaoshan (121 m, 1.35 GW, completed 2003); Jinghong (108 m, 1750 MW, completed 2009); Xiaowan (292 m, 4.2 GW, completed 2010); Gongguoqiao (105 m, 900 MW, completed 2012); Nuozhadu (260 m, 5.85 GW, completed 2014); and Ganlanba (61 m, 155 MW, completed 2019).

In addition, an upper cascade of dams is under development in northern Yunnan Province, consisting of the five dams of Wunonglong,

Lidi, Huangdeng, Dahuaqiao, and Miaowei. Miaowei was the first of the upper cascade dams to complete construction in 2017; the others are scheduled to be in operation by the end of the Thirteenth Five-Year Plan (2020), with two more dams (Gushui and Tuoba) currently in the planning stages. The northernmost segment of the Lancang River, in Yunnan and the Tibet Autonomous Region, is targeted for an additional series of dams, ranging in number from five to eleven, to be built during the Fourteenth Five-Year Plan period (2021–2025) and beyond. Under a policy known as 'Send Western Electricity East', most of Yunnan's hydroelectricity is transported more than 1500 km away to cities in Guangdong Province, where electricity demand in the manufacturing and commercial sectors is high (Magee 2006; Hennig and Magee, this volume).[1]

Social Impacts of Hydropower Projects: Lessons from Early Lancang River Dams

In a major recent review of hydropower development in some of the world's largest river basins, Moran et al. (2018: 11892) conclude that the human costs associated with dams are 'routinely underestimated'. In fact, many of the hydropower projects built around the world over the past century, whether backed by international financial institutions such as the World Bank or by domestic governments, have failed to account for the social disruption they have wrought on the lives of local people. A more accurate understanding of the social and economic impacts of dam-induced displacement has thus become a major focus of research and advocacy for academic institutions, government agencies, international financial institutions, and NGOs around the world (Cernea 2000; Scudder 2005; Richter et al. 2010). Because of the scale and pace of hydropower development in China, dam-induced population resettlement has become a major social concern and has the potential to cause social instability. Despite the importance of dam-induced resettlement, reliable figures are somewhat hard to come by. In 2004, Xinhua News Agency,

[1] The lower reaches of the Mekong in Southeast Asia are also undergoing ambitious hydropower development. Two hydropower projects in Laos (Don Sahong and Xayaburi) are under construction. Nine other projects are in the planning or site preparation stages. China Huaneng Group, one of the country's five major energy generation companies, is playing a major role in these efforts (see Grumbine et al. 2012; Mekong River Commission 2018).

the official media mouthpiece of the Chinese Communist Party, released a short report on research conducted by the Ministry of Water Resources, which concluded that the nation's dam-induced displaced population totalled at least 15 million people, ranking first in the world by a wide margin (Yao 2004). Estimates by researchers, including those at China's National Research Centre for Resettlement, put the total figure at 19.3 million as of 2008 (Shi 2016: 88), and that figure has almost certainly grown by several million in the ensuing years.

A series of recent studies has shown that dam-induced population resettlement in China is associated with a wide range of negative impacts on communities, including reduced land holdings (Chen 2008); reduced access to natural resources and ecological services (Kittinger et al. 2009; Wilmsen et al. 2011); declining household incomes (Tilt et al. 2009; Tilt 2015); widening inter-household and inter-community economic disparities (Wang et al. 2013a); decreased social capital and community connections (Tilt and Gerkey 2016); and diminished mental health and well-being (Xi and Huang 2011). Many of these effects fall disproportionately on vulnerable social groups, including ethnic minorities and low-income communities, who lack the power to participate in political decision-making.

Political leaders and hydropower company representatives in China have recognised these problems since at least the 1950s and sought to address them through a series of policies and directives from the State Council and through various state agencies. Until recently, the most significant policy tool for mitigating the effects of displacement on local populations was an ordinance called 'Regulations on Land-Acquisition Compensation and Resettlement of Migrants for Construction of Large- and Medium-Scale Water Conservancy and Hydropower Projects', which was passed in 1991 and revised by China's State Council in 2006, 2013, and 2017. The 2006 revision greatly increased compensation levels to sixteen times a household's average annual income and also increased compensation for homes and other structures that were inundated or destroyed by a hydropower project (Habich-Sobiegalla and Plümmer, this volume). The ordinance also emphasised that resettled people have a right to know in advance about development plans affecting them and to participate in public hearings, which may give affected people more time to prepare for the effects of resettlement (Tilt 2015). Recognising the inadequacy of resettlement planning and compensation for earlier dam projects, China's central government has retroactively increased compensation levels for villagers at Manwan, the earliest dam on the Lancang River, and other sites (Wang et al. 2013a).

Our household survey, conducted in 2010 in collaboration with American and Chinese researchers, involved 729 households at four of the early hydropower dams completed in the Lancang watershed, including Manwan (1996), Dachaoshan (2003), Xiaowan (2010), and Nuozhadu (2014). We used a cross-sectional sampling approach that compared resettled communities at these dam sites with nearby communities with similar demographic and economic characteristics that had not undergone resettlement. At the time of the survey, tens of thousands of villages had been resettled for these four hydropower projects; currently, the total number of displaced people from all Lancang River dams is estimated at 150,000 (Yunnan Provincial Government 2015).

For displaced households at these four dam sites, the social and economic impacts of resettlement are still unfolding. Because financial compensation is such a key component of resettlement policy and practice, we asked resettled villagers about the amount and type of compensation they had received. When we look at the reported compensation data chronologically, we see gradual improvement over time. In Yun County (the site of Manwan Dam, completed in 1996, and Dachaoshan Dam, completed in 2003), for example, displaced households received an average of 4957 Renminbi (RMB). By contrast, displaced households in Fengqing County (the site of Xiaowan Dam, completed in 2010), received an average of 11,280 RMB, and those in Lancang County (the site of Nuozhadu Dam, under construction at the time of the survey and completed in 2014), received an average of 31,420 RMB.[2] These improvements are directly in line with the State Council ordinance described above that provides for more generous compensation over time. This represents the financial capital that displaced households can use for business investment, educational expenses for children, agricultural supplies, and other means of improving their future economic outlook.

[2] Compensation amounts were adjusted for inflation to 2010 RMB, using a consumer price index from the World Trade Organisation. Because of the political sensitivity surrounding resettlement compensation, only 122 households (of 729 in the survey sample) provided data on compensation amounts. Wang et al. (2013a) report higher compensation rates for resettlers at Manwan Dam, but their figures are based on a different sample. This underscores the difficulty of accurately calculating compensation under shifting policy regimes and amid political sensitivity.

On the other hand, a range of negative social and economic impacts are still being felt by communities displaced for these hydropower projects. Across all of these dam sites, villagers' access to agricultural and forest land was reduced after resettlement, and their agricultural income dropped accordingly; resettled households reported an annual average agricultural income of 15,026 RMB, compared to 19,177 RMB for non-resettled households. In many cases, displaced villagers responded by encouraging family members to find wage employment in cities and towns; resettled households in our survey sample were twice as likely as non-resettled households to report having at least one member of the family working in a city or town and sending remittances back home to supplement household income. In most cases, wage labour or income from self-employment appears to more than make up for the lost agricultural income; the average total annual income was 41,850 RMB for resettled households, compared to 30,565 for their non-resettled counterparts.[3] This appears to be incentivising many resettled villagers to leave the agricultural sector altogether.

Some of the most egregious and long-lasting effects of displacement have played out at the Manwan Dam site, the earliest hydropower project on the Lancang River, where the resettlement scheme failed to adequately plan for and mitigate the effects on local people. Within a few years of the dam's completion in 1996, Chinese researchers began to document widespread poverty and unemployment in the area (Guo 2008). Resettled households also tended to take on significant debt, mostly from family members, as they adjusted to a new living situation and made the necessary investments to get back on their feet.

It is important to note that the effects of resettlement go well beyond the economic and material. For example, social scientists are only beginning to understand how resettlement can affect social capital, the networks of interdependence that people maintain with one another through relationships of trust and reciprocity (Field 2003). For rural villagers in the Lancang River basin, these networks provide a means of

[3] Average household incomes reported in our survey are notably higher than government estimates, which reported annual household incomes in our study counties as approximately 19,000 RMB (Yunnan Statistical Bureau 2011). This is because we calculated income in a more comprehensive way, including agricultural sales minus the costs of inputs; the value of agricultural and livestock production for household consumption; wage labour; self-employment income; and government-provided poverty-alleviation subsidies.

supporting one another by, for example, sharing financial resources or helping one another with farm labour (Tilt and Gerkey 2016). Resettled households in our study showed diminished levels of social capital compared to households that had not undergone resettlement, both in terms of sharing financial resources and sharing farm labour. This represents a reduction in the breadth of networks that farmers rely upon for the labour-intensive tasks of producing rice and other crops for subsistence and market sale. We are also only now beginning to recognise and account for other non-monetary impacts, including the loss of culturally significant sites and effects on emotional and mental well-being.

The 16118 Policy and the Miaowei Dam Case Study

We are now witnessing the implementation of a brand new resettlement policy colloquially known as '16118'. These numbers are abbreviations that stand for specific components of the policy, which we describe in detail in Table 5.1 below (see also GIWP 2009). Formally known as the 'Opinion on Compensation and Subsidies for Resettled People on the Middle Reaches of the Jinsha River for Hydropower Development', the policy was drafted collaboratively by the Yunnan Jinsha River Midstream Hydropower Development Co., Ltd.; various departments of the Yunnan provincial government; and China Hydropower Engineering Consulting Group.

The 16118 policy was adopted by the Yunnan provincial government in 2007 and applied to hydropower projects on the Jinsha River (Upper Yangtze). In 2008, the Yunnan provincial government issued a 'Reply to the issues related to resettlement planning for hydropower development on the Upper Nu River and Upper Lancang River' (Zhu 2018), effectively adopting the standards of the 16118 policy for all hydropower projects in its jurisdiction. It is now becoming the national standard for hydropower resettlement throughout China. By 2011, for example, China's National Development and Reform Commission approved long-term compensation and other provisions of the 16118 policy for displaced people at Sichuan's Lianghekou Hydropower Station. In 2012, the UN Global Compact recognised the 16118 policy with the Best Practice Award for Social Development.

The Miaowei Hydropower Station is located in Gongguoqiao Township, Yunlong County, Dali Prefecture, Yunnan Province. With an installed capacity of 1.4 GW, the facility generates an annual average of

Table 5.1 Key details of the new 16118 resettlement policy

Abbreviation	*Policy Component*	*Details*
1	One Long-Term Compensation Mechanism	Resettled households are compensated based on the average value of crop yields over the three years prior to resettlement. Compensation rates are adjusted periodically by central and provincial government authorities based on local conditions. Households receive remuneration every month for the entire operating period of the hydropower project
6	Simultaneous Implementation of Six Resettlement Methods	Resettled households may be resettled using a variety of mechanisms, singly or in combination, including: (1) resettlement into cities or county towns; (2) resettlement into a combination of urban and rural areas; (3) resettlement into agricultural areas with land provided; (4) dispersed or decentralised resettlement; (5) monetary compensation in addition to, or in lieu of, housing or land; or (6) employment placement
1	Establish One Reservoir Area Industrial Development Fund	After all hydropower projects in a given cascade are developed, a single fund will be established to promote industrial development in the region
1	One Unified Post-Support Policy	Regardless of resettlement method, all resettled households will receive financial support. Specific compensation amounts may be adjusted over time to account for inflation or other economic changes

(continued)

Table 5.1 (continued)

Abbreviation	*Policy Component*	*Details*
8	Eight Resettlement Measures	All hydropower projects will adhere to the following principles: (1) Identify appropriate indicators for compensation and livelihood restoration (e.g. population, land holdings, industrial facilities, cultural sites, etc.); (2) Conduct planning of towns for resettled populations; (3) Implement resettlement by taking account of local conditions; (4) Compensate for rural collective property; (5) Accelerate the transfer of migrant labour into other sectors; (6) Pay attention to social issues and problems in resettlement areas; (7) Help poor households solve subsistence problems; and (8) Ensure environmental protection in the reservoir area

Source Author's creation

6.6 billion kWh of electricity. Completed in 2017, it is the first project in the upstream cascade on the Lancang River to plan and implement population resettlement in accordance with the new 16118 policy.

While the overall changes represented by the 16118 policy are numerous, two major ones stand out. The first is the establishment of long-term compensation rather than a single, one-time payment, which has until recently been the industry standard. Under the 16118 policy, resettled households and their designated descendants or beneficiaries are compensated throughout the entire operational life span of the hydropower project (an uncertain time span, but likely 40–50 years). This represents a recognition of the fact that social and economic consequences for resettled households continue to unfold for a long period of time, as various studies have demonstrated (cf. Tilt and Gerkey 2016). The second major change is a movement away from a one-size-fits-all approach to resettlement in favour of the flexible application of various alternatives that may be more appropriate to different circumstances or attractive for different people (e.g. into cities or towns, into rural areas with agricultural livelihoods, or job retraining). Chinese hydropower company officials have characterised the new policy as a significant milestone in dam-induced displacement and resettlement work (Yunnan Provincial Government 2015). This flexible approach is a movement away from strictly land-based compensation schemes that have long been the industry standard, both for international financial institutions such as the World Bank and the Asian Development Bank and for Chinese hydropower corporations operating under domestic legal statutes and policies (Wilmsen 2018).

Our analysis of the Miaowei Hydropower Project and its social impacts is based on a review of government and hydropower documents and recently published academic studies. In total, 3281 people were resettled into new housing for the project, and an additional 2238 people lost their livelihoods or means of subsistence (e.g. farmland or site of employment). Under China's hydropower resettlement framework, these two groups are known as 'resettled populations' (*banqian anzhi renkou*) and 'production resettlement populations' (*shengchan anzhi renkou*), respectively (Wu 2009).

The affected area included two counties (Yunlong and Lanping), six townships, 21 administrative villages, and 86 agricultural production groups. The resettled population consisted largely of ethnic minorities, primarily Yi and Bai. Because of the mountainous terrain and limited arable land, relocation into agricultural areas was deemed infeasible for most resettlers, particularly since Article 13 of the 16118 policy stipulates that 'resettled people [in agricultural areas] should have agricultural production land equivalent to that of the other residents in the resettlement area' (Zhu 2018: 68). As a result, provincial authorities used four resettlement mechanisms: resettlement into a combination of urban and rural areas; resettlement into agricultural areas with land provided; dispersed resettlement; and monetary compensation.

In addition to direct compensation to affected households, the Miaowei Hydropower Project has involved major infrastructure investment as part of the total project cost, including ten rural housing developments for resettled households, one resettlement townsite, three schools, two medical clinics, three village committee offices, 87 km of roads, two outdoor markets, and two bridges across the Lancang River.

Moreover, several industrial and mining enterprises, and several small-scale hydropower stations, were inundated by the reservoir and rebuilt by project engineers.

Despite positive trends, some challenges remain in the implementation of the 16118 policy. In particular, local authorities had wide discretion over how to enact resettlement plans, and the policy appears to have been implemented quite differently between the two counties in the project area. In Yunlong County, authorities took an approach called 'targeting people, not land' by providing extra annual compensation to households losing more land, paying a portion directly to landholders and another portion to the rural collective. Yunlong County authorities likewise put in place temporary subsidies to displaced households while the resettlement process was taking place, and also paid one-time monetary compensation in two ways: a resettlement compensation fee paid directly to the landholder, and a land compensation fee paid to the rural collective, which in turn was expected to provide various services to displaced people.

In Lanping County, government authorities took the opposite approach of 'targeting land, not people', compensating displaced households based on the annual monetary value of different types of land (paddy fields at 2425 RMB per *mu*, irrigated crop fields at 2834 RMB per *mu*, and non-irrigated fields at 1489 RMB per *mu*). Displaced households in Lanping County did not receive one-time compensation payments. Zhu (2018: 70), studying the Miaowei Hydropower Project, quotes a project engineer who voices these concerns:

> The differences between the two counties in terms of implementation (...) will lead to problems in terms of social stability (...) The resettlement method based on annual compensation has alleviated the pressure on limited land resources, but if the new production and employment opportunities are not enough, this will create more surplus labor and social instability.

At present, very little is known about the experiences and perceptions of people who were resettled for the Miaowei project. Nor is it yet possible to draw conclusions about long-term outcomes for resettlers. Some scholars have expressed scepticism about whether the long-term compensation required under the 16118 policy will prove to be more generous than the previous method of providing one-time compensation based on average annual household agricultural output. Several factors make it difficult to compare compensation at Miaowei with previous projects. First, compensation at Miaowei was calculated for individuals while compensation for earlier projects was calculated for entire households. Second, compensation figures for the earlier projects are based on data reported via a systematic survey, while those for Miaowei are based on published compensation standards, and there is very little available information about how well these standards were actually implemented. Finally, there has yet to be a serious research project examining how the 16118 policy compares with earlier compensation schemes in terms of outcomes for local people.

Nevertheless, as a thought experiment, we attempted to estimate long-term compensation amounts for people displaced by the Miaowei Dam. We estimated total compensation per individual by multiplying the stipulated monthly compensation rate (210 RMB) by the number of months in the compensation period (480 months, or 40 years, a reasonable estimate), which would yield a total compensation figure per individual of 100,800 RMB.

How does this compare with earlier hydropower projects on the Lancang River, which were conducted under the previous resettlement policy regime? Based on our survey data, resettled households at the Nuozhadu Dam site (completed in 2014) received the most generous compensation of the earlier projects, at 31,420 RMB per household. Using an average household size of 3.8 people (Yunnan Statistical Bureau 2011), this would amount to individual compensation of 8268 RMB per person. Thus, monetary compensation standards under the 16118 policy appear to be an order of magnitude more generous than their predecessors. However, these findings should be interpreted with serious caveats. Implementation of the policy, as we have seen in the Miaowei case, is left to county governments, which differ considerably in their capacity to effectively oversee and carry out the provisions of the policy. Furthermore, with a forty-year time horizon for any given project, long-term compensation amounts will be subject to unknown rates of inflation and other market changes. It is also impossible to predict other factors likely to influence the Miaowei Dam's future operations, including future electricity prices, electricity demand, and even the effects of climate change on water availability.

Looking to the Future

Hydropower will likely continue to play an important role in China's energy security and its path towards low-carbon development. However, in the process, it is important to account for the full range of costs associated with dams, including the effects of population displacement and resettlement. Many of these costs remain unexamined and therefore unaddressed.

In an institutional analysis of dam-related resettlement policy in China, Wang et al. (2013b) suggest that policy can be broken down into four chronological eras: 1949–1977 (from the founding of the PRC to the beginning of Reform and Opening); 1978–1993 (the early Reform Period); 1994–2000 (marked by an expanding market and a greater role played by civil society organisations); and 2001–present. Their analysis, published before the 16118 policy was widely known, paints a picture of state policy undergoing fairly gradual evolution. But the new 16118 policy is a major systemic shift; if it continues to be broadly implemented, it will represent a nationwide effort to analyse and mitigate the negative social impacts of dams. We see in this shift reasons for optimism that hydropower resettlement professionals and policy-makers in China have learned some key lessons from the growing body of social science research on this topic, namely that compensation standards must be greatly increased and that long-term support for resettlers must be included in overall project costs.

Looking to the future, a number of trends are worth watching. The first is how resettlement policy will interact with higher legal statutes in China such as the Land Administration Law. In August of 2019, the Standing Committee of the National People's Congress approved new revisions to the law in an effort to clarify the process of land annexation, primarily for large-scale infrastructure projects such as dams and for urban development (Xinhua 2019). These are both highly contentious social issues that regularly provoke protests and demonstrations. It is unclear at present how these revisions will affect hydropower resettlement policies and practices. Depending on how it is implemented, the 16118 policy has the potential to sidestep one of the stickiest problems in rural China: land rights. Owing to the legacy of collective agriculture during the Maoist period, farmers hold long-term use rights but not full ownership of their land. As a result, China's previous resettlement policy regime has been based on a very narrow definition of land use, compensating resettlers primarily based on the agricultural output of inundated farmland, not on its market value or its value for other uses such as the collection of wild resources and forest products. Operating land-based compensation schemes for rural people with limited and insecure land rights has been extremely problematic, as several resettlement experts have pointed out (Wilmsen 2018), and the 16118 policy may therefore represent an improvement.

Another noteworthy trend is the movement towards benefit-sharing for people affected by hydropower projects. The International Hydropower Association, the industry's major global trade group, has developed new benefit-sharing guidelines that suggest a much wider distribution of the benefits of hydropower—including electricity revenue, infrastructure, etc.—with all stakeholders, including resettled people (IHA 2019). The International Hydropower Association (IHA 2019: 7) suggests that benefit-sharing can effectively provide hydropower organisations with what it calls a 'social license to operate' by earning the trust and support of affected people. While benefit-sharing schemes will vary considerably based on local conditions, they would share the following general characteristics: long-term monitoring of social impacts throughout the project life cycle; investment in local infrastructure and livelihood restoration projects; and the provision of equity shares in the hydropower project to local governments and community groups. Chinese resettlement policy is heading in the same direction. With its 2018 'Opinions on Establishing and Perfecting a Hydropower Benefit-Sharing Mechanism', the National Development and Reform Commission signalled its commitment to using project revenues to guarantee living standards for resettlers. To our knowledge, a hydropower project embodying true benefit-sharing does not yet exist anywhere in the world, but researchers and policy-makers are actively trying to envision what the details might look like (Xia et al. 2018). This would effectively recognise resettlers as what the long-time World Bank social scientist and resettlement expert Michael Cernea has called 'investors of equity' who deserve a stake in long-term project benefits (Cernea 2003; see also Cernea 2008).

Finally, it is important to acknowledge that Chinese agencies and firms are taking the lessons learned from hydropower development, much of it in the southwestern region, and applying them globally. In the first two decades of the twenty-first century, Chinese firms have become heavily involved in overseas hydropower development, even as Western-led institutions such as the World Bank have pulled back. This will have implications far beyond China, since Chinese firms are currently involved in the planning, financing, and construction of hundreds of dams in Southeast Asia, Africa, and Latin America (see Sect. 3, this volume). Because Chinese engagement in hydropower development far outstrips that of international financial institutions such as the World Bank, Chinese standards for resettlement and social protections are becoming the world's standards, which means that it is crucial to continually push for greater equity, transparency, and the mitigation of social costs.

Acknowledgments The authors acknowledge funding from the U.S. National Science Foundation, grant 0826752.

References

Cernea, M. (2000). Risks, safeguards, and reconstruction: A model for population displacement and resettlement. In M. Cernea & C. McDowell (Eds.), *Risks and reconstruction: Experience of resettlers and refugees* (pp. 11–55). Washington: World Bank.

Cernea, M. (2003). For a new economics of resettlement: A sociological critique of the compensation principle. *International Social Science Journal, 55*(175), 37–45.

Cernea, M. (2008). Compensation and benefit sharing: Why resettlement policies and practices must be reformed. *Water Science and Engineering, 1*(1), 89–120.

Chen, L. H. (2008). Contradictions in dam building in Yunnan, China: Cultural impacts versus economic growth. *China Report, 44*(2), 97–110.

Field, J. (2003). *Social capital.* New York: Routledge.

GBD Maps Working Group. (2016). *Burden of disease attributable to coal-burning and other major sources of air pollution in China.* Boston: Health Effects Institute.

General Institute of Water Resources and Hydropower Planning and Design (GIWP). (2009). *Specifications on land requisition and resettlement design for construction of water resources and hydropower projects* (SL 290–2009). Beijing: GIWP.

Grumbine, E. R., Dore, J., & Xu, J. C. (2012). Mekong hydropower: Drivers of change and governance challenges. *Frontiers in Ecology and the Environment, 10*(12), 91–98.

Guo, J. J. (2008). *Rethinking development: Anthropological studies on the development of ethnic groups in the Lancang-Mekong River basin.* Kunming: Yunnan People's Press.

International Hydropower Association (IHA). (2019). *How-to guide: Hydropower benefit sharing*. London: IHA.

Kittinger, J. N., Coontz, K. M., Yuan, Z. P., Han, D. J., Zhao, X. F., & Wilcox, B. A. (2009). Toward holistic evaluation and assessment: Linking ecosystems and human well-being for the Three Gorges Dam. *EcoHealth, 6,* 601–613.

Magee, D. (2006). Powershed politics: Hydropower and interprovincial relations under great western development. *China Quarterly, 185,* 23–41.

Mekong River Commission. (2018). *State of the basin report*. Vientiane: Mekong River Commission.

Moran, E. F., Lopez, M. C., Moore, N., Müller, N., & Hyndman, D. W. (2018). Sustainable hydropower in the 21st century. *Proceedings of the National Academy of Sciences, 115*(47), 11891–11898.

Richter, B. D., Postel, S., Revenga, C., Scudder, T., Lehner, B., & Churchill, A. (2010). Lost in development's shadow: The downstream human consequences of dams. *Water Alternatives, 3*(2), 14–42.

Scudder, T. (2005). *The future of large dams: Dealing with social, environmental, institutional and political costs*. London: Earthscan.

Shi, G. Q. (2016). Resettlement induced by dams in China. *Proceedings of the National Research Center for Resettlement*. Nanjing: Hohai University.

Tilt, B. (2015). *Dams and development in China: The moral economy of water and power*. New York: Columbia University Press.

Tilt, B., Braun, Y. A., & He, D. M. (2009). Social impact assessment for large dam projects: A comparison of international projects and implications for best practice. *Journal of Environmental Management, 90*(S3), 249–257.

Tilt, B., & Gerkey, D. (2016). Dams and population displacement on China's Upper Mekong River: Implications for social capital and social-ecological resilience. *Global Environmental Change, 36,* 153–162.

United Nations Development Program (UNDP). (2008). *Human development report: China*. Beijing: UNDP.

Wang, P., Lassoie, J. P., Dong, S. K., & Morreale, S. J. (2013a). A framework for social impact analysis of large dams: A case study of cascading dams on the Upper-Mekong River, China. *Journal of Environmental Management, 117,* 131–140.

Wang, P., Wolf, S. A., Lassoie, J. P., & Dong, S. K. (2013b). Compensation policy for displacement caused by dam construction in China: An institutional analysis. *Geoforum, 48,* 1–9.

Wilmsen, B. (2018). Is land-based resettlement still appropriate for rural people in China? A longitudinal study of displacement at the Three Gorges Dam. *Development and Change, 49*(1), 170–198.

Wilmsen, B., Webber, M., & Duan, Y. F. (2011). Involuntary rural resettlement: Resources, strategies, and outcomes at the Three Gorges Dam, China. *Journal of Environment and Development, 20*(4), 355–380.

Wu, G. S. (2009). Distinction and connection between production resettlement population and Relocation resettlement population. *Guizhou Water Power, 23*(4), 16–19.

Xi, J., & Huang, S. (2011). Relocation stress, coping, and sense of control among resettlers resulting from China's Three Gorges Dam Project. *Social Indicators Research, 104,* 507–522.

Xia, B. Q., Qiang, M. S., Chen, W. C., Fan, Q. X., & Jiang, H. C. (2018). A benefit-sharing model for hydropower projects based on stakeholder input-output analysis: A case study of the Xiluodu project in China. *Land Use Policy, 73,* 341–352.

Xinhua. (2019, August 29). China revises law to clarify land property rights, streamline administration. *XinhuaNet*. http://www.xinhuanet.com/english/2019-08/26/c_138339881.htm. Accessed 1 June 2020.

Yao, R. F. (2004). *Ministry of water resources: China's dam-induced migrants total at least 15 million*. Beijing: Xinhua News.

Yunnan Provincial Government. (2015). *Discussions about resettlement work in the Upper Mekong River Basin*. Kunming.

Yunnan Statistical Bureau. (2011). *Yunnan statistical yearbook*. Kunming: China Statistics Press.

Zhu, J. (2018). Practice and consideration of resettlement with yearly compensation for Miaowei Hydropower Station. *East China Engineering Technology, 143,* 67–71.

CHAPTER 6

Social Stability, Migrant Subjectivities, and Citizenship in China's Resettlement Policies

Sabrina Habich-Sobiegalla and Franziska Plümmer

Introduction

As a result of the Chinese party-state's determination to exploit all possible avenues to ensure China's energy security, instances of dam-induced resettlement have been steadily rising. For several decades now, in order to be able to implement resettlement processes in an orderly manner and to safeguard the party-state's hydropower development strategy, China's resettlement bureaucracy has been designing a regulatory framework for resettlement. These policies increasingly draw on ideas of 'human-oriented' resettlement in an attempt to address the needs

S. Habich-Sobiegalla (✉)
Freie Universität Berlin, Berlin, Germany
e-mail: Sobiegalla@fu-berlin.de

F. Plümmer
University of Vienna, Vienna, Austria

J.-F. Rousseau and S. Habich-Sobiegalla (eds.), *The Political Economy of Hydropower in Southwest China and Beyond*, International Political Economy Series, https://doi.org/10.1007/978-3-030-59361-2_6

of dam migrants rather than merely removing them from sites of future construction and inundation.

This chapter scrutinises the evolution of resettlement schemes through a discussion of citizenship. By looking at the post-resettlement support scheme of 'Constructing a Beautiful Home' (*meili jiayuan jianshe*, MLJY), we show how the state is attempting to construct a feeling of belonging among dam migrant communities so as to give meaning and status to people that have to forsake their homes for a hydropower project. Such support schemes are directed at officially-identified dam migrants, but exclude other project-affected people that have not been subsumed under this category. The government thereby creates a graduated and hierarchical system of migrant citizenships. This assumption builds on Aihwa Ong's (2006) understanding of Chinese migrant identities as governed by technologies of spatial exception, a place-specific regulation differentiating among various groups of citizens and migrants. Accordingly, this chapter investigates the politics of exception that create dam migrant citizenships and subjectivities. Through targeted financial and social rewards accruing to MLJY recipients, dam migrants are made visible for the sake of political propaganda meant to showcase the effectiveness and human-orientedness of resettlement policies. Against this backdrop, this chapter analyses the various technologies applied to target dam migrants collectively and individually, showing that experiences and politics of resettlement, national development, and migration combine to create self-responsible and docile migrant subjects that are proud of their new identity rather than contesting it.

This chapter begins with a theoretical introduction to citizenship practices in Chinese dam migrant communities and briefly introduces the different categories of dam migrants created by resettlement plans. Subsequent sections first describe a recently initiated example of a post-resettlement support programme designed to create a collective dam migrant identity, namely the MLJY scheme. We then illustrate strategies devised by the local government towards individual dam migrants as opposed to the collective. We show that all of these strategies are representative of a resettlement regime that builds on pastoral and benevolent technologies aimed at reducing the perceived risk of social instability and ensuring smooth dam construction processes.

Horses for Courses: Citizenship Practices and Subjectivities in Dam Migrant Communities

To better understand how dam migrants are integrated into the Chinese state project, we build upon Ong's concept of 'graduated citizenship', which describes how the Chinese government applies various governmentalities to its citizens. Ong (2006: 79) argues that 'while low-skilled workers are disciplined, elite workers and members of dominant ethnic groups enjoy affirmative action and pastoral care'. This eventually results in a 'graduated citizenship' according to which the government differentiates how it invests in specific groups of citizens. Similarly, Małgorzata Jakimów (2012: 664) highlights the exclusionary nature of the Chinese concept of citizen (*shimin*), which serves as a 'de facto marker of the boundary between those who "have citizenship rights" and those who do not, and between those who deserve to be identified as "citizens" and those who do not'. Elena Barabantseva (2011: 4f) ascribes this differentiation to the constant references to 'harmonious society' and unity within political speech that have long helped to marginalise the 'internal Other' in order to strengthen the nation state. Against this backdrop, this article scrutinises the technologies used to target dam migrants and their place within the construction of 'a beautiful China'.

The Chinese socialist regime traditionally applies a variety of coercive and non-coercive modalities of government. Sovereignty and discipline have been institutionalised in many spheres, such as social planning through the household registration system (Wang and Liu 2016), birth plans (Greenhalgh and Winckler 2005), and public discourses (Guo 2018), as well as spatial planning (Dutton 1992; Yang 2011). As for dam migrants, we show that state actors apply myriad technologies that determine whether they are marginalised in their local communities. We argue that this governmentality builds on strategies of spatial zoning and social zoning. Spatial zoning describes place-specific technologies collectively aiming at a community that lives in one place and is forced to resettle. Social zoning refers to governmental technologies that target different social groups unevenly.

This approach to social transformation builds on an increasingly sophisticated state apparatus that applies differentiated strategies of 'social engineering for shaping peoples' subjectivities and guiding their conduct from a distance' (Palmer and Winiger 2019: 7). We build on Palmer and

Winiger's understanding of a neo-socialist governmentality that combines three modes of subjectivity within this transformation rationality:

> ...a socialist subjectivity, loyal to the transcendent authority of the Party and trusting its stewardship of the market and of the nation; a market subjectivity, capable of operating autonomously and effectively as producer and consumer in the market economy, and a civilized Chinese subjectivity embodying the excellent virtues and traditions of Chinese civilization and on a par with, if not superior to, liberal Western society. (Palmer and Winiger 2019: 8)

We show that the MLJY scheme contributes to all three subjectivities. First, the project builds on pastoral power, which Michel Foucault defined as power exerted over a flock of people rather than over a specific territory (Foucault 2009: 169). The Chinese Communist Party (CCP) ideologically functions as a shepherd that guides the flock towards solidarity, communal sense, and self-sacrifice. The Party takes the place of the messiah that guides the Chinese nation out of poverty into socialist transformation and improves the lives of all citizens with a promise of salvation. Second, the scheme produces a market subjectivity as it stimulates competition among the different dam migrant communities. Third, MLJY includes a civilising agenda, as it gives clear instructions on how to develop the quality (*suzhi*) of local communities by making them more productive. Moreover, the system rewards and aestheticises the construction of 'traditional' Chinese buildings in dam migrant villages.

This chapter contributes to the literature on neo-socialist governmentality by showing in detail how the MLJY scheme employs discursive and pastoral technologies. In the case of dam migrants, government strategies towards their subjects have changed. In addition, the possibilities of exchanging experiences through social media and online petitioning similarly indicate a shift in state-society interactions, consequently deploying the agency of dam migrants and spurring new forms of government. Moreover, we contribute to debates on migrants' practices of citizenship in the context of resettlement and development (Però 2011; Jakimów 2012; Habich 2015).

Dam(n) Migrants: Changing Rationalities and Technologies in Dam Migrant Regulation

Since the founding of the People's Republic of China in 1949, policies governing dam migrant resettlement have changed significantly. During the Mao era, due to a lack of adequate planning and compensation mechanisms for dam migrants, the living conditions of resettled people were constantly precarious. Only after economic and political reforms were introduced in 1978 did the government begin to introduce measures that would ameliorate the situation. In 1991, state organs published the first sector-specific regulations governing resettlement planning implementation and compensation in the course of water resource projects (State Council 1991). Since their publication, these regulations have been amended twice. In 2006, after having gained resettlement experience through the construction of the Three Gorges Dam as well as through continued cooperation with the World Bank and the Asian Development Bank, the State Council introduced new principles underlying resettlement, requiring that 'resettlement and compensation shall be people-oriented [*yiren weiben*]' (State Council 2006: Section 1, Article 4). In addition, the amount of compensation paid for land was increased from three to four times the average output value of the land to 16 times its average output value during the three years prior to expropriation. In 2017, amended regulations moved away from this standard, stipulating that compensation standards have to be set by provincial governments and have to be equal to the compensation standard set for other infrastructure projects such as roads or railways (State Council 2017).[1]

Apart from resettlement regulations issued by the national government, dam migrants are governed by detailed resettlement plans that are prepared by provincial governments, infrastructure design institutes, and project developers. These plans outline the actual number of dam migrants, the amount of compensation for land crops and houses, and the timeline for resettlement, among other details. According to the resettlement plan for the Nuozhadu Dam, a hydropower station located along

[1] Compensation for infrastructure projects other than water and hydropower projects used to be lower. This policy change reflects the frequent contestations over land compensation by people affected by these other infrastructure projects, who deemed lower compensation standards for the same land as unfair. For further information on resettlement policies, see also the chapters from Wilmsen et al. and Tilt and Chen in this volume.

the Lancang River in Yunnan Province, individuals with permanent residency in the village to be relocated, who *also* legally owned a house, *and* had contracted land with the villager small group would count as dam migrants. In addition, those who had temporarily left the village to serve in the army, go to school, or recent graduates searching for employment were considered to be dam migrants. All others—including permanent military staff, employees of government agencies, and those who had been allocated a job by the state—did not qualify as migrants. Those who had married into a jurisdiction outside the reservoir or construction region were only counted if they had not already changed their household registration (Pu'er Resettlement Bureau 2011).

Alongside the aforementioned change in resettlement policies, the increase of financial compensation, and the formation of increasingly detailed resettlement plans, the treatment of dam migrants by grassroots resettlement officials has changed as well. In line with the requirement to 'guarantee the legitimate rights of resettled people' and undertake 'people-oriented' resettlement (State Council 2006: Article 4), local officials now deploy much more sophisticated techniques of coercion in their attempts to convince villagers to agree to resettlement (Habich 2015). These techniques include so-called participatory rights for migrants, such as the ability to select the type of resettlement (e.g. agricultural or long-term cash support) and the resettlement village, as well as the intensification of village supervision by the local government before and after relocation with propaganda and so-called thought work. The latter refers to local government working groups spending extensive periods of time in villages, speaking to all households individually as well as in town hall meetings, in order to convince everyone that hydropower development is not only in the interest of the nation but also benefits dam migrants; officials sell the move to locals as a unique development opportunity (Pu'er Resettlement Bureau 2011; Habich 2015). Villagers referred to the local government's tactics as a skilful communications strategy of 'cheating, fooling, and threatening (*pian, hong, xia*)' dam migrants (Habich 2015: 178). In one village relocated in the course of the Nuozhadu Dam, for example, villagers were told that their resettlement village would be in a region with beautiful women where single men could find multiple partners. Moreover, ahead of resettlement, local cadres lured villagers with financial payments that were later deducted from their compensation. Local cadres also threatened the use of force if dam migrants refused to accept their offers (Habich 2015).

These strategies stand in contrast to hard forms of coercion deployed in previous decades of dam resettlement, including the forcible removal of villagers from their homes with one week notice by military-trained personnel (Interview, NJ130304). Compared to resettlement processes of the past, current schemes undergo much more rigorous planning and implementation. Nevertheless, such schemes are still conflict-prone, with dam migrants criticising their resettlement compensation as too little and the size of allocated plots of farmland too small to allow farmers to restore their former standards of living. In addition, many resettlement villages are divided over having been resettled at different points in time with different standards of compensation in place. Such uneven standards, and the fact that many migrant households drifted into poverty after resettlement, have caused conflicts between dam migrants and led to protests against local governments (Habich 2016). The next section of this chapter describes a strategy with which the central government has attempted to address these divisions and conflicts.

Meili Jiayuan Jianshe: Pastoral Strategies to Create Collective Migrant Identities

In an article published in January 2013, Tang Chuanli, Director of the Dam Migrant Development Office in the Ministry of Water, explains that the major problems of reservoir-induced resettlement include the striking poverty among dam migrants resettled before 1978, the steep rise in reservoirs constructed in recent years, and a dysfunctional resettlement bureaucracy unable to properly deliver post-resettlement support to dam migrants. Such problems have been identified as hindering the fulfilment of centrally-stipulated tasks such as the construction of a new countryside (*xin nongcun jianshe*), rural-urban integration, fighting poverty, raising rural standards of living, resolving the threefold rural problem (*sannong*), building a beautiful China, and achieving sustainable development of the Chinese nation so as to eventually create a moderately prosperous society (*xiaokang shehui*) by 2020 (Tang 2013). In Yunnan, one of China's poorest provinces, dam migrants have been referred to as 'the weak point' (*duanban*) in the development of a moderately prosperous society; the massive number of dam migrants—expected to reach 1.5 million by 2020, with 20 million others affected by dams—are of special concern to the state (Han 2017).

In order to address these issues, in 2012, the Ministry of Water Resources introduced the MLJY scheme to bring 'blue skies, green fields, and clear water' to all (newly-created and existing) resettlement villages, allowing dam migrants to 'benefit from the achievements of [China's] reform and development' (Tang 2013: online). Depending on local articulation of the central policy, different villages qualify for being turned into a 'beautiful home'. In addition, the particular measures taken in each village depend on its specific needs as defined by the local government. One major focus of MLJY is infrastructure development and the construction or renovation of houses in resettlement villages, as well as the continued resettlement of dam migrants into new resettlement villages. A report by Huize County, Yunnan, is illustrative. Huize is a county with a high number of dam migrants who, according to the local government, are living in poverty. Among the more than 30,000 dam migrants, close to 70% are defined as poor. The county government intends to resolve this problem through the construction of four subtypes of MLJY initiatives, namely: (1) dam migrants fleeing from danger, resolving hardship (*yimin bixian jiekun*); (2) dam migrant new villages (*yimin xincun*); (3) liveable countryside (*yiju xiangcun*); (4) basic infrastructure. While these projects are different in name and target villages, they all entail the construction either of new housing (1–3) or basic infrastructure such as roads, bridges, water and irrigation infrastructure, greenhouses, activity centres, and solar-powered street lights (4) (Nie 2017).

One of the flagship post-resettlement projects implemented in Huize as part of the MLJY scheme is the sub-project of 'dam migrants fleeing from danger, resolving hardship'. This sub-project addresses Huize's dam migrants who had been resettled in the 1960s during the construction of the Maojia Village Dam. These dam migrants have been found to be living in particularly unfortunate conditions with no natural resources or local industries that could support their subsistence. This is why, in 2014, the local government initiated a four-step pilot scheme through which Maojia dam migrants are resettled to new villages and provided with new housing. By 2017, more than 2,000 households had been resettled to newly-created pilot villages in the nearest market town, supposedly close to various natural resources beneficial for dam migrants. Houses have been constructed in Ming/Qing style, which is supposed to represent local architectural characteristics. In order to support dam migrants economically, the local government reports constructing houses that can accommodate commercial activities. In addition, the local government intends to develop agricultural tourism in the region, simultaneously fulfilling the central government's aims to alleviate poverty and revitalise the countryside (Nie 2019). Rousseau and Turner similarly argue that the scheme helps local governments create 'modernised' villagescapes by relocating animal farming from private gardens to collective places or adjusting house design to facilitate cultural tourism (Rousseau and Turner 2018: 144f). Huize is one example in which 'backward' dam migrant communities are to be 'modernised' through resettlement to new villages closer to commercial centres and with sinicised housing. MLJY here not only speaks to the Chinese civilisational subjectivity but also to a market subjectivity.

Another example is Jiangxi Province, where in 2018 county governments started implementing provincial guidelines that required every village with a migrant population of more than 20% to be turned into a MLJY settlement by 2020. This transition entailed improving basic infrastructure, public services, the environment, and the 'democratic and civilised administration' of villages (see Table 6.1). While again most of the improvements come in the form of public goods provided by the local government, some of the stipulations also address villagers themselves. For example, as part of improving the environment, villagers ought to have neat and 'beautiful' (*meiguan*) front porches with all of their possessions stored in an orderly manner. Moreover, a 'democratic and civilised administration' calls on villagers 'to become rich through hard work, to have harmonious relations with their neighbours, to respect the elderly and care for the young, and to be honest and kind' (Anfu County Poverty and Resettlement Office 2018: online). No serious crimes or large protests may occur in resettlement villages.

These guidelines from Jiangxi are in line with other local regulations for the MLJY scheme that can be found in all provinces home to resettlement villages. They illustrate the local governments' hopes to improve village infrastructure and governance. At the same time, these guidelines have a civilising element. This a point made by Director Han from Yunnan, who notes dam migrants' weak 'quality' (*suzhi*). For him the latter refers to the fact that:

> Dam migrants live in remote regions far away from where the market economy develops with inadequate public service provision; their thinking is comparatively backward; their ability to use science and technology and the market for income generation is weak; their overall *suzhi* and abilities

> are low; their traditional ways of production and income generation are hard to change; their production management and income levels are low. (Han 2017: online)

In order to ameliorate this situation, local governments in Yunnan for example offer technical training to dam migrants, which is presented as a way in which dam migrants may 'catch up with the overall level of social development' in China (Liu 2019: online). Here, the government applies a pastoral strategy towards a specific group of citizens that defines clear pathways to become a member of the socialist development project by living orderly and worthy lives, while also embracing a market subjectivity. As such, the MLJY scheme is a guidebook for local communities to improve themselves to (re-)join the Chinese nation. The playbook civilises

Table 6.1 Requirements for MLJY resettlement villages in Jiangxi Province

Type of improvement	*Detailed requirements*
Basic infrastructure	100% of roads paved; clean drinking water and electricity for all households; no open manure pits; 80% of households having harmless and hygienic toilets; public restrooms that are regularly disinfected; television, broadcasting, and internet reception
Public services	Standardised public information boards; daily availability of convenient and quick medical treatment; cultural or sports activity centres
Environment	No burning of waste; drainage and sewage facility coverage of more than 70%; usage of renewable energies by more than 70% of households; clean outer appearance of villages and houses; 35% of villages being 'greenified'
Democratic and civilised administration	Robust grassroots party and village organisations; a functioning democratic system; transparency; guaranteed rights to information; participation and control for all villagers; a village statute; advocating for villagers to become rich through hard work; no serious crimes or large protests

Source Author's creation

not only the front porches of migrants' houses by giving detailed instructions on how they should look; the scheme also targets the 'quality' of the households and the overall community. The resettlement process fosters a new discourse on 'backward' villages and communities that need to lift themselves out of poverty.

Many Chinese local governments, especially those in poverty-stricken areas, are cash-strapped, so the implementation of programmes such as MLJY depends on financial transfers from higher levels of government (Habich-Sobiegalla 2018). In Huize County, for example, 74% of the 252 million Renminbi (RMB) invested in the MLJY scheme has been paid for by upper levels of government (Nie 2017). Many government reports highlight the necessity of attracting upper-level funding, not only to fulfil central policy stipulations, but also to take the opportunity to attract investment to the regions (Liu 2019). In order to attract such funding, governments at the prefectural level and below respond to calls for project applications by provincial governments, detailing their specific plans on how to implement MLJY in their jurisdiction. For example, in 2018, Yunnan's provincial government launched such a competitive call for applications, announcing that 25 of the best projects would be chosen as provincial key projects (*shengji zhongdian xiangmu*). According to the call, the key ingredients for a successful provincial project would be a village of more than 50% dam migrants, of which 'a relatively high number are registered poor households'; where the connection between MLJY and the rural revitalisation campaign is obvious; with a local government and a dam migrant population that is 'active' (*jijixing gao*); and with dam migrants that have a 'uniform will' (*yiyuan tongyi*) (Xie 2018: online). Hence, the provincial government rewards all those villages that fulfil central policy provisions and that have the right kind of attitude towards the project or at least that do not show signs of social conflict. In doing so, the provincial government transfers responsibilities to local governments to optimise their communities. This transfer forges a 'reciprocal responsibility' (Foucault 2009: 225) that makes the local government accountable for the compliance of the migrant community. The local governments, in turn, try to further delegate responsibility to migrants, who are encouraged to be self-responsible and active citizens.

Apart from such prestigious awards by the provincial-level government, local governments can also receive funding outside of such calls. For example, Yuxi, a prefecture-level city in Yunnan, reports that throughout the Thirteenth Five-Year Plan period (2016–2020), the government has received 100 million RMB from upper levels for post-resettlement support, 80% of which went into infrastructure projects. In 2019 alone, the prefecture successfully applied for three new dam migrant village projects with a total investment of 12 million RMB. One MLJY project in Yuxi Prefecture completed in 2017 received three quarters of its funding from the provincial government. The project included water pollution treatment, road pavement, tap water projects, unspecified 'beautification projects', and science, technology, and culture activity centres. The related report states that 'this project turned a poor and backward mountain village into a new socialist village that leans on the mountains and faces the water with a painting-like scenery' (Wu 2017: online).

In sum, the MLJY scheme presents a playbook for creating docile and loyal migrant subjects who work towards beautifying migrant houses, improving their communities, and entrusting the CCP with the stewardship of this project (cf. Palmer and Winiger 2019). The goal of this scheme is to use resettlement as an opportunity to impose a pastoral sense of uplifting Chinese society. In doing so, migrants are encouraged to identify as privileged members of the MLJY scheme. In return, they are expected to become self-responsible Chinese citizens that contribute to building a beautiful China rather than contesting hydropower development and resettlement.

Selective Privileges and Co-optation Strategies

Apart from these strategies to create collective dam migrant identities, local governments have also begun to selectively target certain dam migrants, turning them into agents of the local bureaucracy by making them directly responsible for social stability within their resettlement community. By integrating them into the political organisation, their political citizenship thickens as they become co-responsible for the success of their village's development. For example, in Menglian County, an autonomous county in the southwest of Yunnan Province, three dam migrants act as deputy township governor, deputy village party branch secretary, and assistant village head of the resettlement village. This measure has reportedly been instituted to resolve conflicts between local governments and migrant communities, with migrants themselves acting as intermediaries between the two groups. In another resettlement village that we visited in Yunnan, three village group leaders have been recruited as assistants within government working groups. In return for a monthly payment of 200 RMB each, the assistants support the working group with the implementation of resettlement work and facilitate communication between villagers and cadres. This is not an easy task, as the assistants are exposed to pressure and demands from both fellow dam migrants and the local government. While officials expect the assistants to prevent social instability in this resettlement village where dam migrants were organising protests, the villagers expect their representatives to bring them benefits rather than become part of what they perceive to be an incapable local state.

The recruitment process for such roles in the local government is mostly informal. For example, in one resettlement village, the local government approached the group leader ahead of the Chinese New Year festivities that families in the area celebrate by slaughtering and eating a pig together. Local cadres invited the group leader to this tradition usually practiced among kin, making him part of the 'local state family' as the group leader explained. Reflecting on his role as an intermediary between the resettlement bureaucracy and his fellow villagers, he argued that someone like him was necessary to explain to villagers the importance of hydropower development and resettlement for China and prevent them from continuously 'pestering' the local government (Interview HT130216).

Local governments have also begun to target dam migrants foreseen as particularly troublesome. For example, in Jinggu County, dam migrants who 'pester and make noise' (*chanfang naofang*) are criminalised as part of the nationwide political campaign of 'cracking down on crime, eliminating evil' (*saohei chu'e*) that commenced in early 2018. Depending on the actual implementation of the campaign, criminalised dam migrants might also include those who have been vocal about their rights to adequate compensation and other benefits listed in official resettlement regulation (Pu'er Government 2019). In another prefecture in Yunnan, local officials have singled out households that frequently use the petition system (*shangfang*), co-opting them with a number of unspecified support policies. The local government has thus been able to successfully co-opt former troublemakers, turning them 'into examples of dam migrants that have become rich in the course of the new support system'. This strategy has not just muted critical voices of petitioning households, but has indeed turned them into supporters of the local government who speak favourably of resettlement in front of their peers (Liu 2019: online).

These examples show how local governments meticulously target their strategies towards individual households. If individuals within migrant communities challenge proposed conduct, the official reaction may vary according to how much the government must invest in order to co-opt the subjects. Financially rewarding individual migrants that have been using the petition system is more expensive than organising a symbolic family dinner in order to construct a 'local state family'. The government thus uses benevolence as a strategy to re-integrate possible outliers into the local community.

Conclusion

This chapter has shown how the Chinese government employs benevolent and pastoral strategies towards dam migrants. We contextualised the continuous reforms of dam resettlement policies within a neo-socialist governmentality that applies various strategies of social and spatial zoning. As new methods such as the MLJY scheme seek to integrate local communities into the national development project, we have observed how MLJY differentiates villages and households from each other following a meticulous, often individualised playbook that primarily seeks to maintain social stability. As such, the scheme produces graduated citizenship rights. Migrants are granted a specific place within the social hierarchy according to their loyalty and obedience towards the Party and its local representatives. To ensure compliance, local cadres focus on non-coercive strategies that are geared towards increasing 'population quality' and self-responsible migrants, carefully offering individualised and familial incentives to reel in those migrants who seem less prone to obedience. On a general level, the guidelines of the MLJY programme give detailed instructions on how houses and communities can be improved. This concerns houses and infrastructure as well as the 'population quality' of the dam migrants. The Chinese government upholds the promise of sustainable development and relative welfare for all as a salvation promise that motivates the community to contribute. Migrant communities thus become self-responsible to improve their lives. Outliers who use the petitioning system to complain about local cadres or challenge the fair distribution of compensation funds are kept silent either through punishment or co-optation. As shown above, the strategies applied by local governments to organise resettlement of dam migrants vary, as do levels of compensation. Resettlement programmes have become part China's larger goal to foster socialist transformation specifically by building a purportedly sustainable and modern dam energy network. The goal of the government (both central and local) is to preserve social stability and implement such projects. The central government, specifically the CCP, provides the narrative for delivering a greater good, for bringing the Chinese nation into the future, and for creating a moderately prosperous society. Within this promise of salvation, the local governments act as shepherds guiding migrant communities into a prosperous future through creating socialist subjectivities. These schemes incorporate all levels of government and society, an 'art of government' that operates through assigning (self-)responsibility to each individual.

Migrants' trust in the system is essential for the rationalities informing these strategies to take root. Many of the above-mentioned strategies are symbolic to a certain extent. Without trusting that migrants want to be part of the migrant community, these pastoral strategies would simply not work. Dam migrant identities ascribe meaning to invitations for joint 'family' dinners extended to them by local cadres. Financial compensation, however, also builds on trust—trust that the assigned money was fair and generously distributed. The moral authority of the CCP as the benevolent leader that guides local communities in a modernised society builds on whether the people believe that the rules they live by are enforced. To implement this, local governments must be trusted to transparently forward the funds provided by provincial and central governments and follow their guidelines. In the case of the Nuozhadu Dam in Yunnan Province, however, the villagers could no longer trust the process, as they discovered the communication strategy of 'cheating, fooling, and threatening'. These kinds of non-transparent and patronising strategies contradict the CCP's carefully-built narrative of benevolent guardianship as expressed through the MLJY scheme. Where this image has crumbled, local officials have had to draw on additional strategies targeting individual 'trouble makers' and those who publicly question compensation methods. They additionally draw on soft coercion and co-optation of individual households. Hence, individual privileges are used to co-opt possible outliers and re-integrate them into communities as exceptional examples of how the system works; a collective strategy is thus complemented with individualised strategies. Whether the CCP can regain this trust and continue to implement unequal and opaque policies such as the dam migrant compensation schemes will depend on how the strategies described above effectively prevent public debate and collective action.

References

Anfu County Poverty and Resettlement Office. (2018). *Implementation details on strengthening the construction of beautiful homes for reservoir immigrants*. http://www.afx.gov.cn/xxgk/bmxxgk/fphymb/gsgg_1025/201804/t20180411_701412.htm. Accessed 29 May 2020.

Barabantseva, E. (2011). *Overseas Chinese, ethnic minorities and nationalism: De-centering China*. London: Routledge.

Dutton, M. R. (1992). *Policing and punishment in China. From patriarchy to "the people"*. Cambridge: Cambridge University Press.

Foucault, M. (2009). *Security, territory, population. Lectures at the Collège de France 1977–1978*. Basingstoke: Palgrave Macmillan.

Greenhalgh, S., & Winckler, E. A. (2005). *Governing China's population. From Leninist to neoliberal biopolitics*. Stanford: Stanford University Press.

Guo, B. (2018). *Regulating social media in China. Foucauldian governmentality and the public sphere*. New York: Peter Lang.

Habich, S. (2015). Strategies of soft coercion in Chinese dam resettlement. *Issues and Studies, 51*(1), 165–199.

Habich, S. (2016). *Dams, migration and authoritarianism in China: The local state in Yunnan*. New York: Routledge.

Habich-Sobiegalla, S. (2018). How do central control mechanisms impact local water governance in China? The case of Yunnan Province. *The China Quarterly, 234*, 444–462.

Han, M. (2017). *Thoroughly implement the "four comprehensive" strategy, fully promote the construction of "beautiful homes, moderately prosperous reservoir areas"*. Yunnan Government Resettlement Office. https://archive.org/details/20200529_20200529_0723. Accessed 29 May 2020.

Jakimów, M. (2012). Chinese citizenship 'after orientalism': Academic narratives on internal migrants in China. *Citizenship Studies, 16*(5–6), 657–671.

Liu, H. (2019). "Beautiful homes, moderately prosperous reservoir areas" helps to revitalise countryside. *Yidian Zixun*. https://archive.org/details/20200529_20200529_0636. Accessed 29 May 2020.

Nie, Y. (2017). *The "four projects" in the county of Huize makes whole effort to build "beautiful homes" for migrants*. Yunnan Government Resettlement Office. https://archive.org/details/20200528_20200528_1708. Accessed 29 May 2020.

Nie, Y. (2019). *Huize County eagerly implements the project of dam migrants fleeing from danger, resolving hardship, starting the dam migrant happy life scheme of "beautiful homes, moderately prosperous reservoir areas"*. Yunnan Government Resettlement Office. https://archive.org/details/20200528_20200528_1627. Accessed 29 May 2020.

Ong, A. (2006). *Neoliberalism as exception: Mutations in citizenship and sovereignty*. Durham: Duke University Press.

Palmer, D. A., & Winiger, F. (2019). Neo-socialist governmentality: Managing freedom in the People's Republic of China. *Economy and Society, 48*(4), 554–578.

Però, D. (2011). Migrants' practices of citizenship and policy change. In C. Shore, D. Però, & S. Wright (Eds.), *Policy worlds: Anthropology and the analysis of contemporary power* (pp. 244–265). New York: Berghahn Books.

Pu'er Government. (2019). *The office of relocation and resettlement of the county of Jinggu carries forward the special work on cracking down crimes*. https://archive.org/details/20200529_20200529_0746. Accessed 29 May 2020.

Pu'er Resettlement Bureau. (2011). *Compensation for land appropriation compensation and resettlement in course of Nuozhadu hydropower station construction in Pu'er Municipality*. Handbook for Propagation.

Rousseau, J.-F., & Turner, S. (2018). Not at all costs: Frontier modernization schemes and ethnic minority livelihood debates in the Sino-Vietnamese borderlands. *Verge: Studies in Global Asias, 4*(1),133–159.

State Council. (1991). *Regulations for land appropriation and resettlement induced by large—and medium sized water conservancy and hydropower projects*. State Council Document No. 74.

State Council. (2006). *Regulations for land appropriation and resettlement induced by large—and medium sized water conservancy and hydropower projects*. State Council Document No. 471.

State Council. (2017). *Regulations for land appropriation and resettlement induced by large—and medium sized water conservancy and hydropower projects*. State Council Document No. 679.

Tang, C. (2013). *Vigorously carry out the construction of "beautiful homes" to achieve the noble goal of a moderately prosperous society*. Yunnan Government Resettlement Office. https://archive.org/details/20200528_20200528_1553. Accessed 29 May 2020.

Wang, F., & Liu, Y. (2016). Interpreting Chinese hukou system from a Foucauldian perspective. *Urban Policy and Research, 36*(2), 153–167.

Wu, M. (2017). *Resettlement Bureau of the District of Hongta organises "beautiful homes, moderately prosperous reservoir areas" lake race*. Yunnan Government Resettlement Office. https://archive.org/details/o_20200528. Accessed 29 May 2020.

Xie, Y. (2018). *Provincial resettlement bureau launches competitive project declaration work of 2018 key provincial project "beautiful homes, moderately prosperous reservoir areas" migrant new village*. Yunnan Government Resettlement Office. https://archive.org/details/2018-o. Accessed 29 May 2020.

Yang, J. (2011). The politics of the Dang'an: Spectralization, spatialization, and neoliberal governmentality in China. *Anthropological Quarterly, 84*(2), 507–533.

PART II

Dams and Rural Livelihoods

CHAPTER 7

Green and Pro-Poor? Analysing Social Benefits of Small Hydropower in Yunnan, China

Tyler Harlan

Introduction

Hydropower construction in Yunnan Province has boomed since the 1990s (Hennig et al. 2016). Much of the analysis of the boom has focused on large hydropower (LHP) (Magee 2006; Grumbine et al. 2012; Tilt 2014). Small hydropower (SHP),[1] however, has been mostly ignored, despite making up a third of all installed hydropower capacity in Yunnan (13 GW of 38.2 GW) in 2015 (Hennig et al. 2016). Moreover, installed SHP capacity in Yunnan increased more than fivefold between 1995 and 2015, a rate higher than that of Yunnan's hydropower capacity as a whole (EPS China Data 2020).

[1]There is no agreed-upon global definition of SHP. The International Centre for Small Hydropower (ICSHP) defines SHP as <10 MW; India defines SHP as <25 MW; China defines SHP as <50 MW. Some plants have a dam, and some do not (80% of SHP plants in China do not have a dam; see Cheng 2015).

T. Harlan (✉)
Loyola Marymount University, Los Angeles, CA, USA
e-mail: tyler.harlan@lmu.edu

J.-F. Rousseau and S. Habich-Sobiegalla (eds.), *The Political Economy of Hydropower in Southwest China and Beyond*, International Political Economy Series, https://doi.org/10.1007/978-3-030-59361-2_7

Part of the reason for this gap in analysis is due to the very different functions and impacts of LHP and SHP in Yunnan. The main purpose of LHP has been to generate electricity for load centres in eastern China, and its social and environmental impacts have been severe (Huang and Yan 2009; Tilt and Gerkey 2016; Wilmsen 2016). The main purpose of SHP, by contrast, has historically been to generate electricity for poor rural villagers (Chen 2009; HRC 2009). Since the early 2000s, SHP has also been used to replace peasant fuelwood collection and use with electricity, which officials claim will reduce localised deforestation. SHP would therefore seem to deliver significant benefits for villagers in China, without the negative impacts of LHP. But does it?

This chapter analyses whether or not SHP plants built during the peak decade of the boom (2005–2015) deliver on their green and pro-poor reputation. Nearly all of these plants are situated in grid-connected cascades and have greater installed capacities than plants constructed before the boom. We focus on SHP's role in increasing household electricity use and decreasing household fuelwood collection and use from 2005 to 2015 in Xinping County, Yunnan. We surveyed 122 households in eight village clusters adjacent to SHP plants; we also conducted 62 semi-structured interviews with household heads, plant managers, and county and prefecture government officials. This approach allowed us to pair quantitative data with qualitative detail about specific plants, village contexts, and household behaviours.

Our results show that electricity use and fuelwood collection and use are correlated with the *price* of electricity, which is set by the prefecture government, not the SHP plant. Indeed, SHP plants built during the boom are privately operated and sell electricity directly to the grid, so that they have no say over the price. Most Xinping households therefore report no benefits from SHP. However, a minority of households that receive *subsidised* electricity from one adjacent plant do report benefits from SHP because their electricity is cheaper. This plant is the only one in the county that is sponsored by the central government's 'SHP Replace Fuelwood' (*xiao shuidian dai ranliao*) programme, which provides electricity and construction subsidies; its benefits are limited to adjacent communities. These results reveal that it is only the few subsidised contemporary SHP plants in Yunnan that are green and pro-poor; the vast majority of unsubsidised plants are not.

The chapter proceeds as follows. Section "Green and Pro-Poor SHP?" reviews literature on the benefits and impacts of SHP. Section "Field

Site and Methods" explains our site selection methodology and describes the characteristics of surveyed households. Sections "Electricity Use" and "Fuelwood Collection and Use" analyse and discuss the influence of SHP on household electricity use and household fuelwood collection and use. The final section concludes the chapter.

Green and Pro-Poor SHP?

Traditionally, scholars have viewed SHP as a green, pro-poor technology because of its perceived small environmental footprint and ability to generate reliable electricity for rural households (Bakis and Demirbas 2004; Nautiyal et al. 2011). Much of this work focuses on the use of off-grid SHP systems as a replacement for traditional fuels, such as agricultural residues, fuelwood, and/or charcoal (Mainali and Silveira 2011; Mishra et al. 2015). Studies have shown that burning traditional biomass causes exposure to indoor air pollution (Mishra 2003), is inefficient and time-consuming to collect (Liu et al. 2008), and can exacerbate deforestation (Heltberg et al. 2000). Moreover, research suggests that without electricity, poor households have difficulty raising their incomes because they cannot devote time to employment and education (Cabraal et al. 2005).

Several studies show that SHP can offer a potential way out of this spiral without the high costs associated with grid extension or reliance on diesel generators (Byrne et al. 2007; Urmee et al. 2009). In some cases, poor villagers can also form rural cooperatives and purchase a small-scale system, which they then use to generate electricity for sale to wealthier villagers and pay off the initial loan (Biswas et al. 2001). Scholars recognise that SHP is not a 'silver bullet' solution to poverty, as it must also be affordable and situated within a broader rural development framework. However, it does provide a means to replace traditional fuels, improve productivity, and enable rural people to access services.

Yet recent work also shows that the benefits of SHP can be limited. The major limitation is electricity generation itself. Unlike LHP, SHP plants have little water storage capacity, and thus cannot regulate power supply to meet peak demand (Paish 2002; Bakken et al. 2012). This also means that most SHP plants cannot store water during periods of high rainfall, so excess flow cannot be utilised. This is a problem for off-grid plants because they cannot smooth out fluctuations in generation with electricity imports or exports. Such instability can lead to power cuts and higher electricity prices for SHP-dependent communities (Bakis and Demirbas 2004). Some areas have linked plants together through mini- and rural grids to ensure greater stability, but results are mixed, with some households continuing to use diesel generators and/or fuelwood (Urmee et al. 2009; Mainali and Silveira 2011).

To avoid generation instability, some areas have constructed multiple, larger SHP plants along the same watershed, often (but not always) regulated by a dammed reservoir at the head of the cascade. Due to their inherent connectivity, cascaded SHP projects can more easily respond to changes in rainfall and electricity demand by controlling the output of multiple turbines. Moreover, SHP cascades with a reservoir can continue to operate during dry periods when many diversion-type run-of-river plants must switch off. Such cascaded systems must generally be connected to a regional or national grid, since they produce more electricity than can be consumed locally. The grid operator will typically purchase this electricity at a set rate and sell it back to the local community at a set price.

However, these large grid-connected SHP cascades may offer fewer benefits than off-grid plants. We suggest three main reasons for this based on our findings in this chapter. First, because electricity from large-scale plants is sold to the grid, not to the community, SHP operators have little say over the local electricity price. Plants simply become part of the broader energy infrastructure. Second, because cascaded systems are expensive to build, they are typically operated by profit-minded firms that have little incentive to subsidise electricity or provide local community benefits. Third, once SHP plants are connected to the grid, they must be dispatched alongside (and compete with) other energy sources that are cheaper and more stable, and thus preferred by the grid. As a consequence, SHP may be subject to significant curtailment during periods of low demand or high output, such as the wet season. Such curtailment cuts deeply into profits and resultant tax revenues that could be used for local services.

To date, however, no research has analysed the scenarios identified above, nor how any remaining benefits are distributed between SHP-adjacent communities. Indeed, most studies—including our own previous research in Xinping (Harlan et al. 2020)—emphasise the negative impacts of large-scale grid-connected SHP, which can be significant (Kibler and Tullos 2013; Hennig and Harlan 2018; Kelly 2019). To truly evaluate

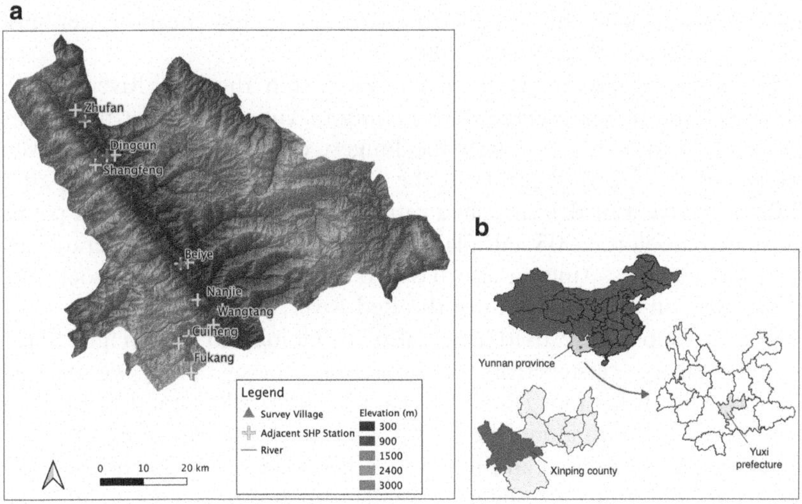

Fig. 7.1 Location of surveyed village clusters and adjacent SHP plants in Xinping County

SHP, though, we must also examine whether or not it delivers the green and pro-poor benefits that it promises. It is to this task that we now turn.

Field Site and Methods

Yunnan has the second-most total installed hydropower capacity and installed SHP capacity of all provinces in China. Xinping is a county in Yunnan's Yuxi Prefecture,[2] situated in the upper reaches of the Red River basin (Fig. 7.1). The county is mountainous, with a total population of 292,000. Average per capita income is 6666 Renminbi (RMB), or 950 US dollars (USD), higher than average for Yunnan due to Xinping's productive agricultural sector. Villages located at upland elevations in Xinping have a lower average per capita income than villages located at lowland elevations. Most lowland villages received their first electricity connection in the early-mid 1970s. This electricity was exclusively generated by SHP and distributed via low-voltage township grids to rural

[2]Yuxi has the administrative rank of prefecture-level city (*shi*) rather than prefecture (*zhou*). We refer to Yuxi as a prefecture for ease of discussion.

households. Upland villages, due to their more remote location, were not fully electrified until the early 2000s.[3]

Xinping contains 26 SHP plants (25 within the Red River basin), which is close to the average for a county in Yunnan. All but five plants are situated in a cascade, with the longest cascade consisting of seven plants. Of the 26 SHP plants in the county, 15 were constructed after 2005 and have a median installed capacity four times greater than plants constructed before 2005. All plants are grid-connected and operated by private companies. Xinping also has two state-owned large hydroelectric dams, both on the main stem of the Red River.

Within Xinping, we used a spatial methodology to select eight SHP-adjacent village clusters for the household survey. First, we divided Xinping's five townships (the administrative unit below county) running from north to south along the Red River into those with more than five plants (three townships) and fewer than five plants (two townships). We chose two sites from each township in the former group and one site from each township in the latter group. Within each township, we subdivided SHP plants into stand-alone and cascade-type systems, larger or smaller than 10 MW installed capacity, and those located at an upland elevation (>800 m) and lowland elevation (<800 m).[4] We then selected plants in each township such that these characteristics were evenly distributed. One of the selected plants is subsidised by the central government through the 'SHP Replace Fuelwood' programme. All but one plant were built during the recent boom. Lastly, we randomly selected 1–3 natural villages[5] located within a single administrative village adjacent to each plant. The number of natural villages selected depended on the density of settlement in the area. Figure 7.1 shows the location of each village cluster and plant. Table 7.1 provides the characteristics of each village cluster and adjacent SHP plant. Village names are pseudonyms.

The survey team consisted of six people split into three pairs. Each pair went door-to-door and surveyed every eighth household; if no one was present, they surveyed the adjacent household. We attempted to interview

[3] Yanzijiao Village, part of the Fukang Village cluster, did not receive an electricity connection until 2004.

[4] In Jianxing Township, which is entirely located above 800 m, we selected one village cluster at 2100 m elevation, and one at 1700 m elevation.

[5] In China, a single village is called a 'natural village,' and a group of natural villages in the same administrative unit is called an 'administrative village.'

Table 7.1 Characteristics of village clusters and adjacent SHP plants

Village cluster	*Elev. [m]*	*Adjacent SHP plant*	*Date plant built*	*Plant installed capacity*	*Subsidised electricity?*
Upland villages					
1. Fukang	2025	Malu	2010	4 MW	Yes
2. Nanjie	835	Nabanqing	2006	10 MW	No
3. Shangfeng	1130	Dachun #1	2008	30 MW	No
4. Cuiheng	1700	Wajiao #3	1991	10 MW	Yes (subsidised by Malu plant)
Lowland villages					
5. Wangtang	666	Wajiao #7	2005	8 MW	No
6. Dingcun	615	Dachun #2	2007	20 MW	No
7. Zhufan	770	Yaocun #2	2013	3.2 MW	No
8. Beiye	551	Nanjian #2	2006	2.5 MW	No

Source Author's creation

residents of both genders and of a range of ages; however, this was largely dependent upon chance. If both a male and female were present, the male tended to answer questions for the household, even if the survey team initially approached the female household member. Of all surveys administered, two (out of 122) are incomplete due to respondents requesting that the interview be terminated.[6] On average, we surveyed 12.5% of households in each village. We coded and transcribed responses after survey administration was complete. Characteristics of surveyed households are provided in Table 7.2.

Electricity Use

Xinping, like rural China more broadly, has experienced significant vacillations in its electricity prices since the early 2000s. Before this period, electricity connections and pricing were the responsibility of township power supply bureaus. Each township in Xinping managed its own minigrid along with one or two SHP stations. Prices for rural household

[6] In both cases, the respondent said that they were busy and needed to return to their work.

Table 7.2 Characteristics of surveyed households

Village cluster	*Total # of households*	*% sampled*	*Mean age of respondent*	*% female respondents*	*Median household income (RMB/mo.)*
Upland villages					
1. Fukang	176	8.5	57	33.3	5500
2. Nanjie	88	12.5	40	27.3	15,250
3. Shangfeng	520	4.0[a]	41	33.3	20,000
4. Cuiheng	186	9.7	55	16.7	15,000
Lowland villages					
5. Wangtang	129	11.6	50	20.0	21,500
6. Dingcun	87	13.8	48	25.0	22,500
7. Zhufan	135	11.9	47	50.0	20,000
8. Beiye	98	14.3	48	0.0	25,000

[a]Shangfeng is larger than the other clusters in our selection because it is a single village where several hundred households were resettled in the past decade. Our sample size is therefore smaller.
Source Author's creation

electricity were set so that the bureaus could recoup costs and re-invest in new infrastructure. These prices tended to be quite high and prohibitive for poor households to use more than a few kilowatt-hours (kWh) for lighting and some agricultural processing. Surveyed households report a wide range of electricity prices during this period, from 0.3 RMB per kWh to 2.5 RMB per kWh, with a mean of 0.62 RMB per kWh.

These prices began to converge in the early 2000s with the establishment of the central government's rural grid renovation programme (*nongwang gaizao*), which aimed to unify urban and rural electricity prices and management by integrating rural power grid infrastructure with prefectural, provincial, and national grids. At the national level, the government invested 290 billion RMB to renovate grid infrastructure, which decreased wire losses of rural grids from 25–30% to less than 12% (Peng and Pan 2006: 82). This investment was accompanied by reforms to electricity management that transformed local power supply bureaus into wholly-owned subsidiaries of provincial grid companies (which themselves are subsidiaries of the regional grid company). The electricity price within this institutional structure is set by the Yuxi Prefecture government, not the township government.

As a result of this programme, electricity prices in Xinping have been de-coupled from the local electricity infrastructure. Even as Xinping added new SHP plants after 2005, the electricity price was set by the Yuxi government. Grid renovation in Xinping was 85% complete in 2015 and has reduced the electricity price for residents of surveyed villages to an average of 0.46 RMB per kWh ($N = 74$). This price is the same for nearly all households, regardless of which SHP plant they receive electricity from. Thus, once villages are fully integrated into the national grid infrastructure, they are no longer affected by a single SHP plant or township grid.

There is, however, one major exception to this rule. In the early 2000s, at the same time as rural grid renovation, the central government instituted the 'SHP Replace Fuelwood' programme. This programme aimed to reduce deforestation resulting from fuelwood collection in the upper watersheds of major rivers by subsidising electricity in poor, upland villages (Tang et al. 2012). This programme provides two subsidies: (1) a subsidy to the SHP developer covering half of all construction costs, and (2) a subsidy to households in the target area that reduces the electricity price by 50%. This second subsidy may also be combined with the provision of discounted appliances (such as televisions and refrigerators) to households to encourage electricity use. SHP plants that are not part of the 'SHP Replace Fuelwood' programme are not subsidised.

Of the 25 SHP plants in Xinping, only one is subsidised by the 'SHP Replace Fuelwood' programme. This plant, called Malu River SHP, has 4 MW of installed capacity and generates electricity for industrial and residential use in Jianxing Township, which includes the surveyed upland villages of Fukang and Cuiheng. Jianxing Township was chosen for the programme because it is poor and considered to have a serious problem with deforestation (Interview with Xinping State Forestry Administration official). Residential users in Jianxing receive an electricity subsidy of 0.22 RMB per kWh, a reduction of 50% of the original price. All other residential electricity users in Xinping pay the 'normal' price of approximately 0.46 RMB per kWh.

Given this reduction in electricity prices after grid renovation, we would expect a corresponding increase in electricity use across all surveyed villages, regardless of which SHP plant their electricity is generated from. For Fukang and Cuiheng Villages, which have subsidised electricity, we would also expect increased usage. Use changes by village are displayed in Fig. 7.2. Since grid renovation, rural electricity consumption has risen in all surveyed villages, with an average increase of 52.8 kWh per month per household ($N = 42$). Some 89% of surveyed households report a

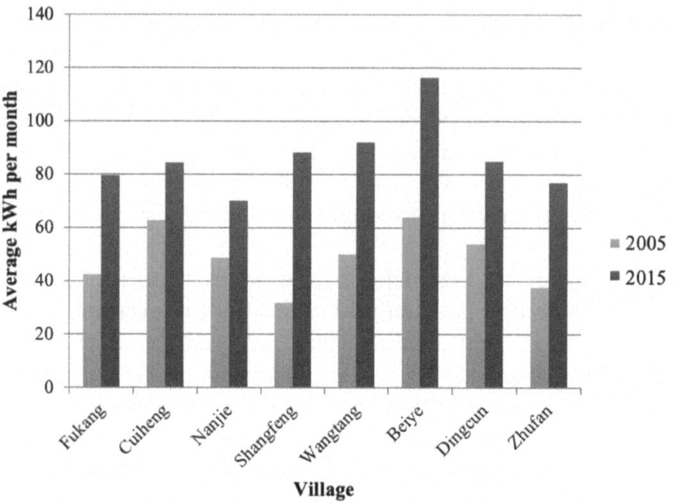

Fig. 7.2 Average electricity use before/after grid renovation (2005 and 2015) (*Note* For 2005, $N = 48$. For 2015, $N = 96$)

change in electricity use since grid renovation between 2005 and 2015 ($N = 102$).

In interviews, nearly all households cited the electricity price as a key determinant of their increased usage. However, only households in subsidised villages describe their increased usage as a benefit of the SHP plant. This difference between unsubsidised and subsidised villages is borne out in responses to an open-ended question about the benefits of SHP. In unsubsidised villages, only ten households (11.4%, $N = 89$) reported positive electricity benefits, and four of these households had received subsidies from a plant because their land had been used for SHP construction. This answer is unsurprising given that SHP plants have no effect on electricity prices in these villages. In Fukang, however, 33% ($N = 15$) of households stated that electricity use is cheaper and/or more convenient because of the SHP plant. In Cuiheng, 50% of households ($N = 18$) stated the same. Here, we can conclude that positive perceptions about the electricity benefits of SHP are mainly confined to areas subsidised by the 'SHP Replace Fuelwood' project. Thus, SHP is only pro-poor when it is subsidised.

Fuelwood Collection and Use

In addition to promoting SHP for rural electrification—despite evidence to the contrary—the government also claims that SHP is a 'green' technology that reduces peasant reliance on fuelwood collection. Officials blame fuelwood collection for localised deforestation, which can cause erosion on steep hillsides and exacerbate flooding and poor water quality in watersheds. Collecting fuelwood is also time-consuming, difficult work that reduces households' available labour time for other income-generating activities. Moreover, burning fuelwood in poorly ventilated homes can cause respiratory illness. Our survey team conducted several interviews in the dark, mud-brick homes of upland villages while our hosts added wood to an open fire to keep warm. Even after a few minutes, our eyes and throats would burn from smoke inhalation. In contrast, the modern brick-and-tile homes of many lowland villagers had dedicated areas for burning wood, which was generally only done during large celebrations. Here, using electricity instead of fuelwood has clear health and livelihood benefits.

In Xinping, the increase in electricity use described above is mirrored by a decrease in fuelwood use since 2005. In the survey, 99.2% of households said that they used fuelwood in 2005, while 75.4% responded that they used fuelwood in 2015. Figure 7.3 shows the average amount of

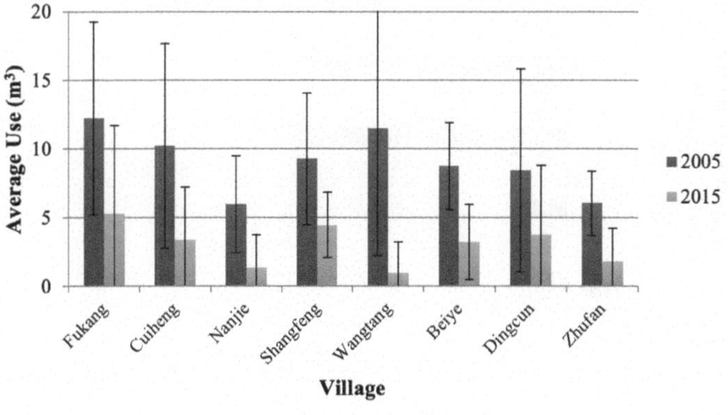

Fig. 7.3 Average fuelwood use, 2005 and 2015 (*Note* For both categories, $N = 122$)

fuelwood used by each village. Note that even with high standard deviations within each village sample, fuelwood use in 2015 is much less than in 2005 across all villages. The 'SHP Replace Fuelwood' villages of Fukang and Cuiheng also show a reduction in fuelwood use, even as they started from a slightly higher average base. That these subsidised villages display a similar trend to others is unsurprising, since the electricity subsidy has simply smoothed out differences that we might otherwise see.

Moreover, households that still used fuelwood in 2015 changed why they do so. In the decades before grid renovation, households in Xinping used fuelwood mainly for heating and cooking. Those with livestock also used fuelwood to warm pig feed. However, with the reduction in electricity prices, households are now more willing to replace fuelwood with electric heaters and cooking appliances. Fuelwood use for heating and cooking have both reduced by approximately 50%, and households that burn fuelwood for cooking report that it is mostly for boiling water or at big events like weddings or new year celebrations. Changes to the types of fuelwood use between 2005 and 2015 are shown in Fig. 7.4.

Let us examine fuelwood for heating and cooking in more detail, as these two uses are proxies for dependence upon burning biomass for

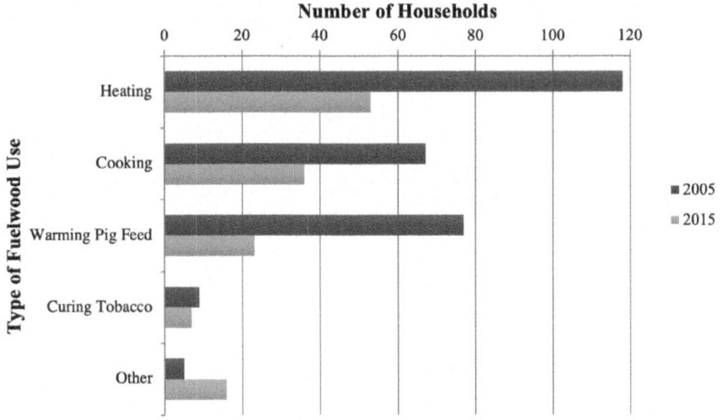

Fig. 7.4 Types of fuelwood use, 2005 and 2015 (*Note* For both categories, $N = 122$)

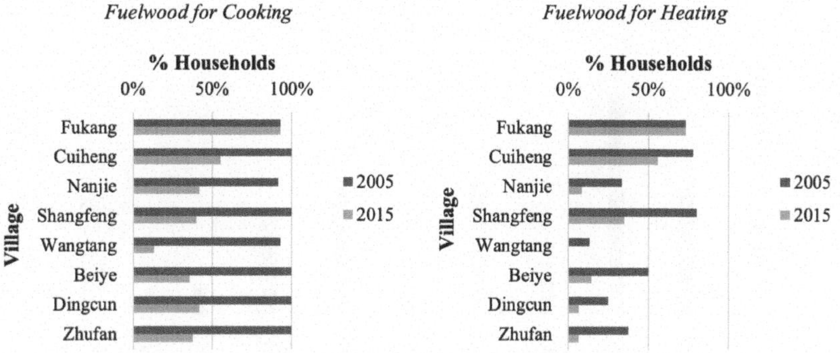

Fig. 7.5 Household fuelwood use for cooking (left) and heating (right) (*Note* For all categories, $N = 120$)

daily activities.[7] The percentage of households in each village that used fuelwood for cooking and heating is shown in Fig. 7.5. Households in Fukang and Cuiheng used fuelwood for cooking and heating in 2015 more than households in other villages.

Figure 7.5 contradicts the story depicted in Fig. 7.3, which shows a reduction in fuelwood use across all surveyed villages. Why is it that burning fuelwood for cooking and heating persisted in 2015 in over half of the households in Fukang and Cuiheng? The main reason is because some poor households in these villages cannot afford to use electricity to meet all their cooking and heating needs, even though their electricity is subsidised. In qualitative interviews, several respondents in Fukang and Cuiheng said that they supplement electric appliances with fuelwood, especially on cold nights. The 'SHP Replace Fuelwood' programme has reduced fuelwood use for these activities in the two villages, but it has not eliminated use altogether. This suggests that the subsidy has been well targeted to villages in Xinping that are the most reliant on fuelwood and stand to benefit the most from its reduction. Nonetheless, subsidised electricity cannot fully reduce dependence on fuelwood for the poorest households.

[7]This is more the case for heating than for cooking, since 'cooking' may include the few meals per year that coincide with large celebrations.

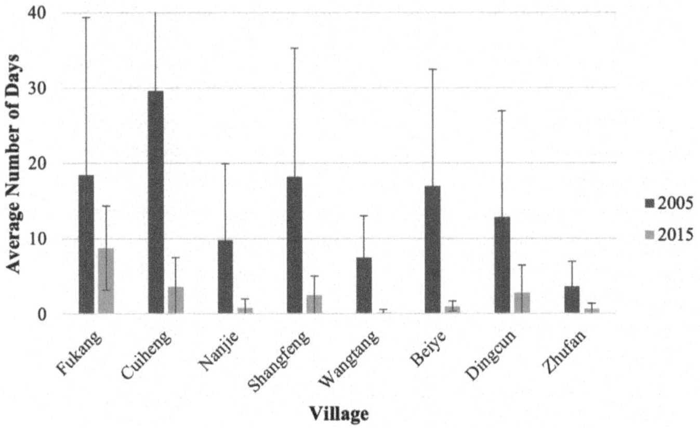

Fig. 7.6 Number of days collecting fuelwood, 2005 and 2015 (*Note* For 2005, $N = 84$. For 2015, $N = 90$)

Finally, beyond fuelwood use alone, we must also examine changes to labour time required for its collection. In 2005, before rural grid renovation in Xinping, surveyed households reported spending an average of 15.7 person-days per year collecting wood ($N = 84$).[8] By 2015, the average had fallen to 2.6 person-days per year ($N = 90$). Figure 7.6 shows changes to time spent collecting fuelwood by village between 2005 and 2015.

From this figure, we can see that the average amount of time spent collecting fuelwood varies between villages. In 2005, households in Cuiheng village spent an average of 29.6 person-days collecting wood, while the villages of Fukang, Shangfeng, and Beiye spent approximately 18 person-days. All other villages expended significantly less time collecting wood in 2005. By 2015, all villages reported a significant reduction, with only Fukang (8.7 person-days), Cuiheng (3.7), Dingcun (2.8), and Shangfeng (2.5) spending more than one day of labour-time collecting wood. Here again, we can interpret this persistence of fuelwood collection as a function of poverty, with households collecting wood to supplement electricity use.

[8] One 'person-day' is equivalent to eight hours of labour.

That the subsidised villages of Fukang and Cuiheng reduced fuelwood collection by 88% and 53%, respectively, suggests that the 'SHP Replace Fuelwood' programme has had some positive effect on household livelihoods. Nonetheless, we must place these gains in the broader context of SHP expansion in Xinping. As argued above, it is only the 'SHP Replace Fuelwood' villages of Fukang and Cuiheng that receive subsidised electricity from an SHP plant. Villages outside this zone pay the 'normal' electricity price, which is set by the Yuxi government. Thus, while the county-wide reduction in fuelwood collection and use is a positive outcome, it is mainly the result of rural grid renovation and electricity reforms, not SHP. Programmes like 'SHP Replace Fuelwood' that subsidise electricity for poor villages can induce households to reduce fuelwood collection and use beyond what might occur otherwise.

Conclusion

The Chinese government has long promoted SHP as a green, pro-poor technology in mountainous areas of Southwest China like Xinping. And, indeed, in the years before the 2000s SHP boom and grid renovation, many households in Southwest China received electricity from SHP. Since the boom, however, SHP plants have become larger, situated in cascades, and fully grid-connected, with little incentive to provide local benefits. This chapter has analysed the reduction and distribution of social benefits of SHP when such a shift in type of SHP occurs.

Our survey of households in Xinping found that electricity use and fuelwood collection and use are unaffected by SHP, except for villages subsidised by the central government's 'SHP Replace Fuelwood' project. In the early 2000s, much of rural China, including Xinping, underwent grid renovation and electricity management reforms that normalised the price of electricity, which severed the ability for SHP plants to set a sale price. As a result, any increase in electricity consumption in Xinping, and corresponding drop in fuelwood collection and use, cannot be tied to SHP. Rather, these reductions are a result of fewer family members remaining in the village, rising incomes (for some households), and a cheaper electricity price after grid renovation. It is only the villages of Fukang and Cuiheng, which receive subsidised electricity, that report benefits from the SHP plant. Thus, while the 'SHP Replace Fuelwood' plants produced some positive impacts, all other plants in Xinping built during the boom produced no positive impacts. We must therefore be sceptical of claims that large, grid-connected SHP improves livelihoods and local forest cover in Yunnan; in most cases, it does not.

The findings of this chapter must also be placed in the broader context of the social impacts of SHP in Yunnan, including negative impacts. As we show in previous work (Harlan 2018; Hennig and Harlan 2018; Harlan et al. 2020), villages in Xinping that are situated between a plant's intake and outflow are highly susceptible to reduced irrigation water access during the dry season. Water access issues are further exacerbated by the installed capacity of the adjacent plant, the quality of village irrigation channels, and whether or not plant operators release water during dry periods. In some cases, households are forced to change their crops to water-saving varieties or construct water-saving irrigation systems, both of which are expensive and disruptive to livelihoods. Moreover, this exposes households to higher capital costs and cash crop price swings, which are increasingly burdensome as smallholder farming in China becomes less and less profitable. Thus, in Yunnan, it is *only* subsidised SHP plants that that can be described as green and pro-poor, while most others are the opposite.

The findings of this chapter raise further questions in need of analysis. In particular, given the small sample size of surveyed households, it is unclear how effective electricity subsidies have been in reducing fuelwood collection and use (and therefore deforestation) in Xinping and other parts of Southwest China. While surveyed residents report benefiting from subsidies and have reduced fuelwood collection and use, these changes could also be driven by other factors. A large-sample survey conducted across a broader geographical scale would enable researchers to better elucidate this relationship. Such a survey could also be paired with remote sensing data to compare forest cover change in 'SHP Replace Fuelwood' zones with change in unsubsidised areas.

SHP is not a green, pro-poor technology in and of itself; it is shaped by the conditions in which it is managed and subsidised. Across the world, SHP plants have provided steady electricity to communities who previously had no access, including tens of millions of households in China. In certain times and places, SHP can be a green, pro-poor driver of rural development, and well-sited and well-managed plants should continue to be deployed in areas lacking electricity and grid infrastructure. In contemporary China, however, SHP has ridden on its past coattails for too long. Unsubsidised plants may provide valuable revenue to local governments—and potentially replace polluting activities—but they are not the green rural electrification sources that they are made out to be. Future expansion of SHP thus must be limited to carefully sited and managed plants that have tangible livelihood benefits.

Acknowledgements The author acknowledges funding from the US State Department (Fulbright-IIE Program); the American Council of Learned Societies; the UCLA Asia Institute; and the UCLA Center for Chinese Studies.

References

Bakis, R., & Demirbas, A. (2004). Sustainable development of small hydropower plants (SHPs). *Energy Sources, 26*(12), 1105–1118.

Bakken, T. H., Sundt, H., Ruud, A., & Harby, A. (2012). Development of small versus large hydropower in Norway: Comparison of environmental impacts. *Energy Procedia, 20,* 185–199.

Biswas, W. K., Bryce, P., & Diesendorf, M. (2001). Model for empowering rural poor through renewable energy technologies in Bangladesh. *Environmental Science & Policy, 4*(6), 333–344.

Byrne, J., Zhou, A., Shen, B., & Hughes, K. (2007). Evaluating the potential of small-scale renewable energy options to meet rural livelihoods needs: A GIS- and lifecycle cost-based assessment of Western China's options. *Energy Policy, 35*(8), 4391–4401.

Cabraal, R. A., Barnes, D. F., & Agarwal, S. G. (2005). Productive uses of energy for rural development. *Annual Review of Environment and Resources, 30*(1), 117–144.

Chen, J. (2009). Develop small hydropower and promote rural economic development. *China Hydropower and Electrification, 5,* 10–11.

Cheng, X. (2015). *A general survey of SHP development in China.* Presented at the Hangzhou Regional Center for Small Hydropower, Hangzhou.

EPS China Data. (2020). *China water conservancy statistics.* http://www.epschinadata.com/. Accessed 1 June 2020.

Grumbine, R. E., Dore, J., & Xu, J. (2012). Mekong hydropower: Drivers of change and governance challenges. *Frontiers in Ecology and the Environment, 10*(2), 91–98.

Hangzhou Regional Centre for Small Hydropower (HRC). (2009). *Rural hydropower and electrification in China* (2nd ed.). Hangzhou: China Water Press.

Harlan, T. (2018). Rural utility to low-carbon industry: Small hydropower and the industrialization of renewable energy in China. *Geoforum, 95,* 59–69.

Harlan, T., Xu, R., & He, J. (2020). Is small hydropower beautiful? Social impacts of river fragmentation in China's Red River Basin. *Ambio,* Online First, 1–12.

Heltberg, R., Arndt, T. C., & Sekhar, N. U. (2000). Fuelwood consumption and forest degradation: A household model for domestic energy substitution in rural India. *Land Economics, 76*(2), 213–232.

Hennig, T., & Harlan, T. (2018). Shades of green energy: Geographies of small hydropower in Yunnan, China and the challenges of over-development. *Global Environmental Change, 49,* 116–128.

Hennig, T., Wang, W., Magee, D., & He, D. (2016). Yunnan's fast-paced large hydropower development: A powershed-based approach to critically assessing generation and consumption paradigms. *Water, 8*(10), 1–24.

Huang, H., & Yan, Z. (2009). Present situation and future prospect of hydropower in China. *Renewable and Sustainable Energy Reviews, 13*(6–7), 1652–1656.

Kelly, S. (2019). Megawatts mask impacts: Small hydropower and knowledge politics in the Puelwillimapu, Southern Chile. *Energy Research & Social Science, 54,* 224–235.

Kibler, K. M., & Tullos, D. D. (2013). Cumulative biophysical impact of small and large hydropower development in Nu River, China. *Water Resources Research, 49,* 3104–3118.

Liu, G., Lucas, M., & Shen, L. (2008). Rural household energy consumption and its impacts on eco-environment in Tibet: Taking Taktse county as an example. *Renewable and Sustainable Energy Reviews, 12*(7), 1890–1908.

Magee, D. (2006). Powershed politics: Yunnan hydropower under Great Western Development. *The China Quarterly, 185,* 23–41.

Mainali, B., & Silveira, S. (2011). Financing off-grid rural electrification: Country case Nepal. *5th Dubrovnik Conference on Sustainable Development of Energy, Water & Environment Systems, 36*(4), 2194–2201.

Mishra, M. K., Khare, N., & Agrawal, A. B. (2015). Small hydro power in India: Current status and future perspectives. *Renewable and Sustainable Energy Reviews, 51,* 101–115.

Mishra, V. (2003). Indoor air pollution from biomass combustion and acute respiratory illness in preschool age children in Zimbabwe. *International Journal of Epidemiology, 32*(5), 847–853.

Nautiyal, H., Singal, S. K., Goel, V., & Sharma, A. (2011). Small hydropower for sustainable energy development in India. *Renewable and Sustainable Energy Reviews, 15*(4), 2021–2027.

Paish, O. (2002). Small hydro power: Technology and current status. *Renewable and Sustainable Energy Reviews, 6*(6), 537–556.

Peng, W., & Pan, J. (2006). Rural electrification in China: History and institution. *China & World Economy, 14*(1), 71–84.

Tang, X., Li, Q., Wu, M., Tang, W., Jin, F., Haynes, J., ... (2012). Ecological environment protection in Chinese rural hydropower development practices: A review. *Water, Air, & Soil Pollution, 223*(6), 3033–3048.

Tilt, B. (2014). *Dams and development in China: The moral economy of water and power*. New York: Columbia University Press.

Tilt, B., & Gerkey, D. (2016). Dams and population displacement on China's Upper Mekong River: Implications for social capital and social–ecological resilience. *Global Environmental Change, 36*, 153–162.

Urmee, T., Harries, D., & Schlapfer, A. (2009). Issues related to rural electrification using renewable energy in developing countries of Asia and Pacific. *Renewable Energy, 34*(2), 354–357.

Wilmsen, B. (2016). After the deluge: A longitudinal study of resettlement at the Three Gorges Dam, China. *World Development, 84*, 41–54.

CHAPTER 8

Small Hydropower for Electricity and Modernity: Impacts on the Everyday Lives of Minority Communities in Yunnan's Nu River Valley

Thomas Ptak

Introduction

While a range of modernisation phenomena have dramatically transformed much of present-day China, many rural areas have been excluded and face increasing marginalisation. Contemporary research investigating China's transformation has centred largely on economic developments since liberalisation began in 1978. Regions across the eastern seaboard, Pearl River Delta, and urban cores such as Beijing, Chongqing, Shanghai, and Guangzhou are disproportionately represented in studies (see, for example, Wu and Phelps 2011; Lim 2014; Wu 2015). Consequently, rural and remote areas, particularly in Southwest China (Ptak et al. 2019),

T. Ptak (✉)
University of Idaho, Moscow, ID, USA
e-mail: tptak@uidaho.edu

J.-F. Rousseau and S. Habich-Sobiegalla (eds.), *The Political Economy of Hydropower in Southwest China and Beyond*, International Political Economy Series, https://doi.org/10.1007/978-3-030-59361-2_8

have not received adequate attention, a trend indicative of broader spatial imbalances.

One area that has received significant attention, however, is the contemporary proliferation of hydroelectricity, given the key role it plays in China's energy landscape and modernisation drive. There is a well-established body of literature analysing hydropower development in China through topics such as environmental sustainability (Kibbler and Tullos 2013), state-society relations (Habich 2016; Rousseau 2019), development (Ptak 2014; Tilt 2014), and broader issues related to electricity generation and exchange (Hennig et al. 2016; Ptak and Hommel 2016; Liu et al. 2018). There is a dearth of critical research, however, investigating how hydropower is used as a tool for modernising social circumstances, in this case through rural electrification. In order to provide fresh insights into this understudied phenomenon, the present chapter critically investigates how hydropower-based rural electrification programmes—framed by local governments through discourses of modernity—have influenced the daily lives of ethnic minority communities, particularly those in remote areas struggling in the context of China's dramatic transformation. The following question reflects the core of this research: How has small hydropower-based electrification influenced the social circumstances and everyday lives of residents in Dimaluo, a small, remote ethnic minority community in the upper Nu River Valley?

This study responds to Woods' (2007) call for scholars to continue researching rural areas, where the forces of modernity and globalisation are often highly evident, yet often overlooked due to the allure of hypermodern urban centres. The key objective of this research is to reveal and develop a more comprehensive and nuanced understanding of direct and unintended outcomes of rural electrification programmes. Specifically, the goal is to shed light on the ways local communities navigate and negotiate modernity and to determine to what extent globalisation acts as a homogenising force in shaping daily practices, local culture, and identity (Boli and Petova 2007; Ritzer 2011) within a broader context of dynamic social change across China's rural countryside (Oakes 1998; Liu 2004).

Drawing on empirical data and insights developed while living among small rural communities in Yunnan, this research advances limited studies of the upper Nu River Valley in far northwestern Yunnan Province. Specifically, this chapter casts new light on prior work by Harwood (2013), who studied how rural livelihood changes, and labour practices in particular, reflected processes of modernisation. Harwood posited that residents of the upper Nu River Valley are becoming increasingly intertwined with the development agendas of China's central authorities and the expanding market economy. The case study presented in this chapter uses Harwood's work as a lens but focuses on everyday lifestyle changes rather than solely labour practices to determine outcomes in relation to Harwood's findings.

China's Grand Canyon of the East

High up in a remote corner of northwestern Yunnan Province, the Nu River Valley dominates the Nujiang Autonomous Prefecture (Fig. 8.1). Extending for over three hundred km and known as China's 'Grand Canyon of the East' (Goodman 2010), the valley comprises one section of the UNESCO-designated Three Parallel Rivers of Yunnan World Heritage Areas. Defined by the Gaoligongshan Mountains along its western flank and Biluoxueshan (Biluo Snow Mountains) to the east, the valley averages a depth of 2000 m. These mountain ranges extend up to 4600 m in height and influence climatic conditions both in the valley and the broader Three Parallel Rivers ecoregion.

Remote and isolated, the valley represents one of China's most unique biological zones; its 70% forest cover rate is one of the highest in the country. According to UNESCO (2012: online), during the last ice age, the region acted as a biological refuge and has been described as 'one of the world's least disturbed temperate ecological areas, an epicentre of Chinese endemic species and a natural gene pool of great richness'. While the Nu River Valley contains vast biological characteristics, it also exhibits a wealth of cultural diversity, as ethnic minority groups make up the vast majority of its residents. In the upper Nu River Valley, 92.2% of residents in Gongshan Dulong and Nu Autonomous County are ethnic minorities, officially known as *shaoshu minzu*. Tibetans, Lisu, Nu, Pumi, Yi, and Bai reside in the valley. The high percentage of ethnic minority communities complicates many challenges linked to underdevelopment in the region (Barabantseva 2009). Differences in core belief structures, class status, and history underpin relations among minority groups and between minorities and Han Chinese, who occupy the overwhelming majority of government positions (Ptak 2019).

While abundant in natural resources, the three counties of the Nujiang Autonomous Prefecture located in the Nu River Valley (Gongshan–Dulong–Nu, Fugong, and Lushui) are among the poorest in China. Even

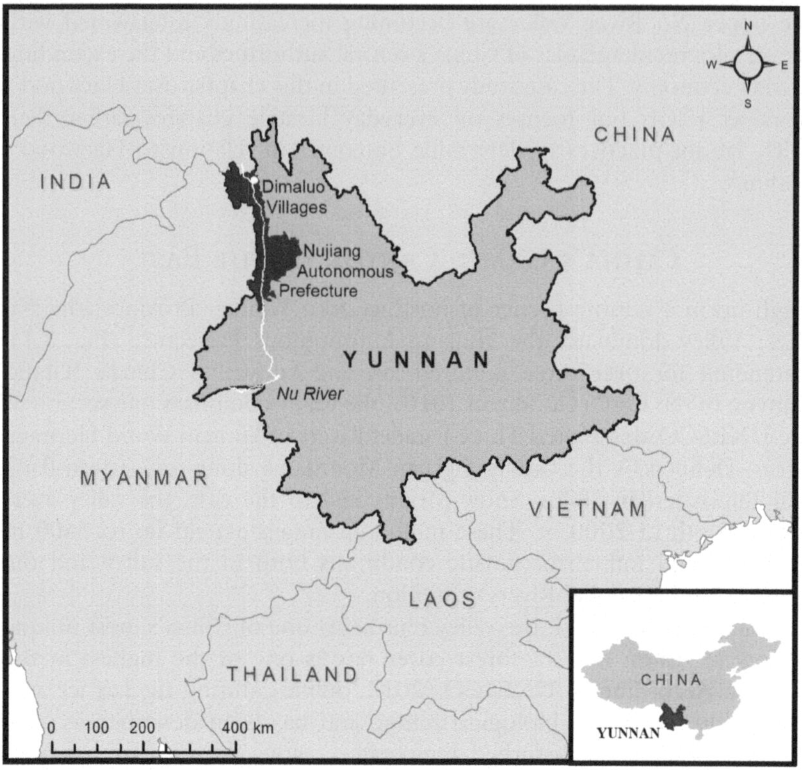

Fig. 8.1 Map of Yunnan and the Nujiang Autonomous Prefecture (*Source* Author)

for Yunnan Province, which remains near the bottom of China's wealth indices, the Nu River Valley has low per capita incomes. In 2016, average per capita disposable income for households in Yunnan was 16,719 Renminbi (RMB), with an urban average of 28,610 RMB and a rural average of 9020 RMB (China Statistical Yearbook 2019). In 2016, the average disposable income for households in the Nujiang Autonomous Prefecture was 5299 RMB—less than half of the provincial average for rural households and less than a third of the overall provincial average (Statistical Bureau of Yunnan Province 2017).

The Villages and Community of Dimaluo

The boundaries of Yunnan, Myanmar, and the Tibetan Autonomous Region converge at the northern end of the Nujiang Autonomous Prefecture. Approximately halfway between the administrative capital of Gongshan and the small township of Bingzhongluo, a nondescript road leaves the main stem of the Nu River and heads up into the Biluo Mountains. After five km, a community of villages known collectively as Dimaluo appears. Dimaluo is a community of approximately 3000 residents living in 12 separate villages situated on the steep valley and floodplains adjoining the Dimaluo River, a tributary of the Nu. The Dimaluo Valley was originally inhabited by Nu people before Tibetans began migrating over the Biluo Mountains from Deqin County in the Lancang River Valley. While the word 'Dimaluo' itself is Lisu in its origin, the majority of the community's population are ethnically Nu and Tibetan. The villages' dominant language and cultural traits (such as singing, dancing, and dress style) are also Tibetan. Villagers typically engage in small-scale subsistence agricultural practices in addition to animal husbandry. Fields of canola, peppers, onions, and cabbage dominate the landscape of the Dimaluo Valley, while corn, the most common crop, provides feed for pigs and chickens. Some villagers own cows and yaks that reside in higher alpine meadows.

The community of Dimaluo offers a unique opportunity to undertake a study into the efficacy of rural electrification programmes, as it contains some of the last villages in Yunnan to receive electricity. The impact (or lack thereof) of electrification on everyday life is easier to measure when compared to communities who have received electricity for a substantial period of time. In order to develop a holistic understanding of the local community and obtain sufficient data, it was necessary to spend significant time engaged in field research. Consequently, the study was undertaken through a mixed-methods ethnographic approach. Between 2012 and 2018, I made five separate trips to the villages of Dimaluo and spent a total of twelve months living among the community. This staggered approach to field research allowed for critical reflexivity (Dowling 2000), with qualitative data obtained through a suite of participant observation techniques (Evans 1988; Kearns 2005), household surveys, semi-structured interviews, and focus groups. Quantitative datasets were obtained from the websites of the Chinese Bureau of Statistics, Nujiang Autonomous Prefecture Government, and the Gongshan Dulong and

Nu Autonomous County Government, along with data from the Yunnan Statistical Yearbook.

Everyday Life in Dimaluo Prior to Electricity

In order to understand how electrification has shaped the everyday lives of residents in Dimaluo, it is necessary to understand conditions prior to 1997, when the floodplains below Dimaluo's main village of Qiumatong were first identified as suitable for hosting a small hydroelectric dam. In early 1998, construction began on a road to Dimaluo. The road was crudely blasted out of cliffs and carved into land along the path of the Dimaluo River.

Residents repeatedly related how life prior to the road and electricity was exceedingly difficult for the community as a whole. In order to illuminate these previous social circumstances, two key quotes summarise the broader narrative communicated by local residents regarding the marginal nature of everyday life in Dimaluo. Many local residents described the struggle simply to put food on their tables. An older man with whom I spent time each day remembered particular hardships, such as the quality and availability of food:

> Before the dam was built life was very difficult. We were never comfortable and could never relax. We only ever ate corn that we grew; it took us a day to walk to the Nu River Valley to buy rice, and then a day to walk back. We never had nice clothes and jackets like we have now. All of the food that we planted we had to eat, we could never sell any to make income. Now we have the market in Dimaluo because of the road and the Chinese people we can trade our goods with. It is much better now, we can relax a little bit and enjoy our life. Also, since the dam was built, we have a road – we can go all the way to Gongshan in one hour. (Interview 68, March 2014)

The word *xinkule* was repeated often during conversations about life before road access to the valley. *Xinkule* denotes hardship or hard work, and a younger man illuminated some of the hardships experienced before the road:

> It was 1997, before the road was built, and when we got it, we had a huge celebration. Before the road I would go to school in Pandang. Life was the most difficult, it was *xinkule*, because we had to walk and it would

> take one day. Also, we had to pay for school and books, but we had no money. Back then the food was very bad, there was not even rice, only corn flour and some vegetables like carrots and cabbage. Every month you would only eat meat like pork once or twice, now we eat it every day. We were so poor back then and our clothes were like rags. There was no support from the government. We didn't have any concrete and when it rained everything was mud. At night there were no lights, so we would sit in the dark unless we had candles, which were expensive, or the fireplace, which made so much smoke. (Interview 7, November 2013)

When local residents were first told of plans to build the dam, many were apprehensive about the changes expected to sweep through their isolated and underdeveloped but largely peaceful valley. Some believed economic opportunities and the chance to move away from the daily rigours of farming would accompany construction of the dam and new road. Many residents had been informed by local government officials that their daily lives would improve through electrical supply, and for some—but not all—this turned out to be true. Narratives from local government officials implored the need to improve their low-quality state and become more modern. Traditional ways of thinking and acting, they said, must change. In China, *di suzhi* is the term used when speaking of low quality in relation to humans. Although there is no direct translation in English, *di suzhi* refers broadly to human development. The term denotes segments of China's society—not overtly, but implicitly, ethnic minority groups—who are viewed as possessing low levels of sophistication while contributing minimal economic and cultural value to the Chinese state (Jacka 2009). It must be noted, however, that the term *di suzhi* is not used in reference towards China's ethnic minorities only.

Everyday Life in Dimaluo After Electrification

Construction on the small pumped-storage embankment dam first began in 2003 and was completed in 2009. In the same year, the generating station fitted with two 12 Mw turbines began producing hydroelectricity. Although some villages in the community had already started to receive electricity in advance of the dam's completion, the majority had not. On my first visit to Dimaluo in 2012, it was evident that electrification of the valley was still ongoing. The town square in Dimaluo's main village of Qiumatong contained power lines, transformer boxes, and other infrastructure necessary to expand the nascent electrical network (Fig. 8.2). Initial electrification and ongoing expansion of the network was developed under the auspices of a Rural Electrification Programme (REP).

Since the founding of the People's Republic of China in 1949, central, provincial, and local governments have developed and implemented multiple initiatives aimed at delivering electricity to rural areas. The range of initiatives and programmes collectively fall under the REP banner. Recently, small hydropower has been prioritised as an effective tool for providing electricity in rural and remote mountainous areas where grid penetration is difficult or impractical (Zhang and Kumar 2011; see also Harlan, this volume). The Nu River Valley REP has focused on development of small hydropower in order to provide electricity for rural and remote villages (Luo and Guo 2013; Zhaohong and Yanling 2015; Hennig and Magee, this volume).

In order to evaluate whether the everyday lives of residents had changed since the small dam was developed and the local community

Fig. 8.2 Electrical infrastructure in Dimaluo's main village (*Source* Author)

was provided with electricity, it was necessary to determine if there were substantial changes from subsistence farming to wage generating forms of labour as a direct result of electrification. Qualitative data from household surveys, semi-structured interviews, and focus groups showed that there had not been substantial changes in daily occupation for the vast majority of residents in each of Dimaluo's 12 villages. While a few residents had been able to leverage electricity to establish wage generating occupations, 98% of local residents interviewed and surveyed still engaged in subsistence agriculture on an everyday basis. Clearly, electrification itself was not a determinant for significantly altering daily labour-based practices for most residents.

Access to electricity also did not influence everyday practices pertaining to heating in any of the households surveyed. During fieldwork in Dimaluo, much time was spent inside the houses of local residents, and fires were omnipresent. When discussing the use of fires with one respondent, he detailed the complex relationship the community has with fire as a daily social practice:

> Well you see, it doesn't work with our houses to make heat with electricity. The walls are open and the roof is open, so we would need concrete houses if that was going to work. I think many people care about the fire smoke in their house – it is dirty, and for many people this is bad. We should have cleaner air to breathe, and inside some of the houses it is so black. But after some time you get used to it. And many people smoke cigarettes and drink a lot, so the smoke from the fire is not so bad. It is really bad for the children though, but what can we do, we are poor and this is how we must live. But you have to understand the fire is a social place, we have sat around the fire for many years and that is where we talk and socialise. (Interview 72, April 2014)

While electricity did not meaningfully impact heating or labour-based practices across Dimaluo, it was significant in shaping daily practices of residents through electrical appliances. Household survey data illuminated how two main appliances, rice cookers and hot plates, were present in 100% of dwellings visited. Prior to electrification, it was only possible to cook using fires which were located on the floor of the main room in dwellings (Fig. 8.3). Although most residents still used a combination of electrical appliances and fire depending on what was being prepared (e.g. the local corn alcohol was prepared over a long slow fire for which

Fig. 8.3 Prior to electrification cooking was only possible through fires (*Source* Author)

appliances were not suitable), data overwhelmingly showed that appliances reduced the amount of time one was required to spend inside each day.

One respondent described how cooking with electrical appliances and the use of refrigeration (present in only 5% of dwellings surveyed) had meaningfully shaped their daily life:

> Using electricity to cook is very good; we use it make rice so fast and mostly to fry vegetables. Before we would always cook rice and vegetables on the fireplace, but now we prefer the new way because it is so easy and saves us time. We also have a fridge that we can store food in. Before we could not do this so it is very convenient. When we would kill our animals we had to eat the meat immediately, or preserve them with salt. We could only do it in the winter time. Now we can store the food in the fridge and it doesn't matter. Also we can make cheese and sometimes butter to sell at

> the market. Without the fridge they would go bad very quickly. (Interview 89, August 2014)

Similarly to cooking with electrical appliances, washing machines (present in 66% of dwellings surveyed) saved residents valuable time when compared to hand washing clothing. Time is a precious asset for residents living in rural and marginal agricultural communities. Therefore, freeing up time each day allowed people to concentrate on other tasks, such as gathering feed for livestock, tilling fields, or collecting wood to heat dwellings or cook. Furthermore, it was common for respondents to clearly explain the connection between freeing of time through electrical supply and increases in what they termed daily productivity or opportunities to enjoy life a little more. The following quote details this scenario:

> For me, when I think about all of the things that electricity has brought, I like the washing machine the best. The washing machine saves me so much time every day. Washing clothes by hand is very difficult because first I need to collect so much water and carry it back from the river. Then I have to heat the water on the fire and fill up the wash tub, then I wash the clothes and finally rinse them and hang them to dry. In the winter this is very difficult. If I have many clothes, I must use more than one tub of water, and it takes such a long time. Also, it is very hard on my hands. Now with the washing machine it is very easy, everything is done at once and I can use the time I save to work on other important things like collecting wood or mushrooms, cooking, feeding the pigs, or making corn wine which I can sell to make some money. However, I also use the time I save to relax a little (laughs). (Interview 14, November 2013)

While cooking appliances and washing machines significantly shaped everyday activities and can be understood as a reflection of modernity (they were only available since the valley was electrified), televisions (identified in 84% of dwellings) likewise impacted the everyday lives of residents in substantial ways. First, televisions were a medium for local residents to obtain previously unavailable information. Many remarked on enjoying the daily news transmitted from the local station and the national CCTV network (available for those with a satellite dish) for understanding events across the country and the world. Televisions were also used for relaxation in the evenings. During one interview a resident explained:

> The television is so good – we really like this. Many people have them and we enjoy it because at night after all our work is done we can watch it and relax, we couldn't do this before the dam and electricity. Televisions are very important because we can get lots of information about local politics, and also the young kids can study Chinese just by watching the television. We also can see different places around the world and learn about new things. (Interview 24, November 2013)

Television also influenced daily activities in Dimaluo by shaping new forms of recreation. Basketball represents the most popular sport in China. In the upper Nu River Valley, particularly the villages of Dimaluo, it was clearly a favourite pastime. Even remote villages high up on the valley walls contained basketball courts. During conversations with residents, it became apparent that the ability of local people to watch America's National Basketball Association (NBA) games through satellite television had established and then enhanced the popularity of the game.

Basketball's popularity was shaped by the Chinese mega-star Yao Ming, whose career coincided with the first televisions arriving in Dimaluo. It was not uncommon to see young men wearing NBA jerseys around the villages, and during morning game broadcasts, people sat huddled around televisions watching their favourite teams. Most afternoons, local people would gather to socialise while teams played games on the court (Fig. 8.4). Many local residents communicated their love for playing and watching basketball and professed their appreciation of electricity for making it possible.

Electrification also made possible two other technologies that significantly shaped the everyday lives of Dimaluo residents, namely the internet and cellular phones. Cell phones are ubiquitous throughout Dimaluo. Data obtained during fieldwork and in informal discussions with villagers showed that 99% of local residents owned cell phones, with younger residents also connecting to the internet via their smartphones. Use of cellular technology is pervasive across the contemporary world, particularly in China, which boasts the largest number of cell phone users globally at around one billion—approximately 79% of the country's population (Statista 2019).

Data obtained through interviews and participant observation with younger residents demonstrated a 100% connection rate to the Internet through smartphone use. Furthermore, younger residents demonstrated a 100% connection rate to China's most popular social media application,

Fig. 8.4 Local residents playing basketball in Dimaluo (*Source* Author)

known as WeChat in English. It was very common to see young people in Dimaluo speaking into their cell phones, sending photos or videos, and keeping up with current events (Fig. 8.5).

Many residents explained how they used their smartphones to watch pop music videos, following artists from Korea just as much as Tibetan or Chinese artists. This is important to note, as traditional Tibetan music, particularly folk songs played on an instrument known as the *xianzi*, is popular among younger residents in Dimaluo. While they enjoy modern music and other influences through web-based applications, Dimaluo's younger generation have not discarded traditional modes of culture as a consequence of modernity. The experiences of these youth challenge notions of globalisation as a force homogenising daily practices, culture, and identities (Boli and Petova 2007; Ritzer 2011). The above data also support prior work from Liu (2004) and Oakes (1998), who illuminate the nuances and complexities characterising how small, rural

Fig. 8.5 A local resident connecting to the internet via a smartphone (*Source* Author)

villages navigate and negotiate dynamic social changes shaped by forces of globalisation and modernity.

Cell phones also created a time-space compression in the Dimaluo Valley, as they negated the need for many residents to hike for days across steep, often treacherous mountain paths just to communicate with friends or family in remote villages. Without a local electrical supply to support the cellular network, these types of interactions would not be possible.

One day while I was talking with a group of residents, an elderly respondent described how cell phone technology had influenced his daily life. With a laugh, he also outlined concerns that cell phones were becoming too important for some members of the community:

> I've only had the small phone for a few years, but I really like it. We can keep in touch with friends and don't have to walk to town to talk with them, although I do like to do this. But if the weather is bad we can stay home and talk. Having the small phone is good when someone is sick. We can call a car to come and get us – before it was very hard as we had to have someone go and find us a horse, but now it's so easy. All of the younger kids have the phone with a computer on it. I think they spend too much time playing games or watching movies on their phones. All the time they are always on their phone and this is bad. The computer phones are important, but you cannot spend your whole life on the phone! (Interview 112, September 2017)

Unintended Consequences of Electrification on Everyday Lives

There can be a tendency to focus on the overt dimensions of hydropower development when conducting research, as the most evident outcomes are also the easiest to measure. However, ethnographic research allows us to look beyond explicit empirical outcomes and discover more surreptitious elements that may only be revealed through time and/or multiple methodologies. Yet even at an early stage of my stay in the Dimaluo Valley, it became apparent that two major unintended consequences of dam construction and electrification had substantially influenced the everyday lives of local residents. First, construction of the road shaped positive change; this link fundamentally influenced the daily practices of most residents by providing them with opportunities for mobility, something previously limited to walking paths. Second was the adverse influence of a specific commodity, beer, which prior to the road had been difficult to transport to the communities.

The road connecting Dimaluo to the Nu River Valley, necessary for the construction of the dam, was a quantum leap for many residents as it opened up the community to the benefits of vehicular transportation, a fundamental prerequisite to modernity. Initially, only a single road connected the Nu River Valley to the main village of Qiumatong.

However, the road advanced incrementally along the main stem of the Dimaluo River. After a number of years, additional roads were built connecting smaller and more remote villages located on the valley walls. Finally, in 2015, a road reached the last village of Alou La Ka, and all of Dimaluo was now accessible by vehicle. Furthermore, in 2017, the main road extended up and over the Biluo mountains to the Mekong River Valley, creating the first road connection between the upper Nu River and Mekong River Valleys.

Mobility brought the most significant changes to everyday life in the community. Prior to the road there was no weekly market in the town square of Qiumatong Village. Since then, however, local traders from villages of the upper Nu River Valley and Han Chinese merchants from as far as Baoshan (420 km away), arrive each Thursday to sell a range of goods, including processed food, clothing, health products, tools, and electronics. Residents of Dimaluo also use the weekly market to sell agricultural products; as it represents the best opportunity for a wide range of local people to generate income, several have been able to change their occupation and become full-time traders. Furthermore, residents are now able to move freely between their villages and other regions in the Nu River Valley. In particular, people often travel to the administrative capital of Gongshan to sell goods, seek medical care, engage in wage labour, or undertake administrative tasks.

Enhanced mobility also offered opportunities for some people to establish businesses and move away from subsistence agriculture. As there are no public busses serving Dimaluo, transportation is only possible through informal networks. The most common mode of transportation are minivans, known across China as *mianbaoche*, which translates to 'bread loaf van'. Local residents owned a small fleet of minivans and shuttled people back and forth between Dimaluo and various locations each day (Fig. 8.6). So, although not a direct result of electrification itself, accompanying infrastructure (in this case the road) has significantly shaped the daily lives of local residents.

Prior to the construction of the road, local people relied on the production and consumption of a local corn-based alcohol known as meat wine or *xiala*. As corn was also the staple crop utilised as stock feed for animals and food for humans, *xiala* production was limited. With the completion of the road, small stores emerged and were immediately filled with beer. As beer became readily available, consumption

Fig. 8.6 A *mianbaoche* loading up with cargo and local residents (*Source* Author)

rapidly increased. The consequences of an exponential increase in alcohol consumption had a dramatic influence on the small community.

One evening, I sat with a local resident who explained how the availability of large quantities of beer had adversely changed social circumstances in the community:

> It is very sad because you can see so many people who are completely addicted to alcohol. When the road was built all of the beer became available and we could see so many people drinking all the time. Before, it was only the alcohol we could make, and only at special times; now it is every day. The younger people grow up seeing everyone drinking all the time, then they start to drink early. It's very sad and there is a big problem with this. Also, if you walk around the villages, you can see so much rubbish. People throw their beer bottles into nature or into the river. This is very bad. (Interview 81, July 2014)

Fig. 8.7 Discarded beer bottles in Dimaluo (*Source* Author)

Moreover, he explained that a decision had been made by the local leadership to ban the sale of beer in bottles in village stores. Bottles, as opposed to cans, could not be recycled and consequently ended up being discarded (Fig. 8.7), creating environmental challenges.

Thus, the road's opening up of Dimaluo, necessary for electrification through construction of the dam, spurred an spike in alcohol consumption by the local community. The availability of beer, transported to the villages via the new network of roads, has transformed social circumstances in the community as alcohol, once reserved for celebrations or special occasions, has become a major part of everyday life.

Discussion

The villages of Dimaluo have undergone a gradual transformation from previously unelectrified, isolated, remote settlements into a community with tangible (albeit limited) connections to processes linked to China's drive to modernity. As this case study demonstrates, hydroelectricity itself has not significantly altered social circumstances and everyday practices for most residents in Dimaluo. However, hydroelectricity generated by the small dam and developed through the REP has directly introduced electrical appliances, televisions, cell phones, and the Internet along with indirect outcomes such as mobility, market access, and alcohol consumption.

These research findings contrast with those of Harwood (2013), who argued that residents of Yunnan's upper Nu River Valley will become increasingly intertwined with the development agendas of China's central authorities and expanding market economy. Harwood attributes rural modernisation and opportunities for improving social circumstances to national government reforms in agriculture and the development of macro-scale infrastructure such as roads and railways. He argues that time and space between China's peripheral ethnic minority regions and the market-based growth engines of the eastern seaboard have collapsed. While this case study indeed demonstrates that electrification drove a time-space compression, I argue that this process has not significantly connected the Dimaluo community to market forces in ways that have been transformational.

There are two principal rationales for explaining the differences between Harwood's study and this research. First, Harwood conducted his research along the main stem of the Nu River Valley, mainly in and around the urban and administrative centre of Gongshan. Dimaluo's location, removed from the main commercial and transportation networks in the Nu River Valley, shapes the types of possible and practical connections regardless of globalisation. Geography clearly matters, and this case study demonstrates the importance of considering differences between Dimaluo and Gongshan—which are only 20 km apart—as well as the differences between China's underdeveloped periphery and its core. Furthermore, Harwood's study investigated labour and education and focused on individuals, while this research analyses the impacts of hydroelectricity development on the broader social circumstances in which whole communities are shaped.

Dimaluo can be seen as a microcosm of the broader ongoing challenges faced by ethnic minority communities across China. At the heart of these challenges lie the relationships between China's core and its periphery: that is, eastern, urban centres and Han Chinese versus rural, remote regions and ethnic minorities. In these relationships, connectivity from the periphery to the core is often weak and largely based on dependency. In contrast, connectivity from the core to the periphery is strong, defined by the core's ongoing development utilising resources, energy, and labour at the expense of peripheral, already marginalised areas.

He (2007) argues that this situation reflects China's ongoing struggle to integrate poor, rural residents into national development. The problem is located in the rapid penetration of consumer culture into rural areas. Televisions, along with cellular and Internet technology, are central to this shift, and as this chapter has demonstrated, even the most remote and isolated residents are now exposed to consumerism. Their limited incomes, however, make it impossible for them to fulfil all their consumption desires. Consequently, many of China's poor rural residents are unable to change their status or begin to engage in modern, consumer-based practices. They thus become further marginalised. Their confidence—already low—is further weakened due to shame, as a pervasive consumer culture informs them that local knowledge is wrong, traditional beliefs are ignorant, customary lifestyle objectives are uninteresting, and ultimately they are still of low quality (*di suzhi*).

Conclusions

This chapter has addressed the following research question: How has small hydropower-based electrification influenced the social circumstances and everyday lives of residents in Dimaluo, a small, remote ethnic minority community in the upper Nu River Valley? My findings support those of Mackerras (2003), who has argued that the physical isolation of ethnic minority communities in Yunnan will continue restricting their ability to engage with modernity and to meaningfully change their social circumstances, regardless of technological advances. Sixteen years after Mackerras's study was published, wealth inequalities and economic marginalisation continue to increase, further exposing spatial disparities between core and peripheral regions in Yunnan and across China more broadly. The simultaneous deepening of processes linked to modernity for some and a lack of substantial changes to daily practices and everyday life for many others have increased the feeling of isolation. In this case study, electrification as a stand-alone strategy has been shown to be ineffective as a tool for creating meaningful social change and substantially influencing the daily practices of ethnic minority residents.

Electricity, and indeed all forms of energy, are critical to any modernisation project. Energy, however, can only provide baseline conditions and its limitations must also be understood. In underdeveloped, rural, and remote ethnic minority regions, electrification cannot simply be viewed as a panacea. As recent research has shown (Boamah and Rothfuß 2018; Kumar 2018; Malakar 2018), if rural electrification is to deliver meaningful transformation, it must be introduced strategically as one component within a broader suite of complementary programmes. Otherwise, China's rural and remote regions, particularly those inhabited by ethnic minority groups, will be further marginalised. It is not modernisation per se or a particular type of modernisation that marginalises China's minority residents. Marginalisation often occurs when modernisation projects are imposed upon communities through programmes such as rural electrification without benefit-sharing mechanisms (Jie et al. 2013). A more holistic approach to rural electrification programmes must be adopted, and it ought to consider a wider range of social complexities, particularly when it comes to small hydropower projects in rural and remote ethnic minority areas (Klimpt et al. 2002). More inclusive approaches will ensure that electrification provides a larger percentage of rural communities with the ability to overcome poverty and change social circumstances. If new approaches are not implemented, however, many of China's ethnic minorities and rural residents will remain in the dichotomous position of being connected, but largely obscured, by shadows cast upon them through modernisation and globalisation.

References

Barabantseva, V. (2009). Development as localization, ethnic minorities in China's official discourse on the western development project. *Critical Asian Studies, 41*(2), 225–254.

Boamah, F., & Rothfuß, E. (2018). From technical innovations towards social practices and socio-technical transition? Re-thinking the transition to decentralised solar PV electrification in Africa. *Energy Research and Social Science, 42,* 1–10.

Boli, J., & Petrova, V. (2007). Globalization today. In G. Ritzer (Ed.), *The Blackwell companion to globalization* (pp. 103–124). Malden: Blackwell.

China Statistical Yearbook. (2019). *Per capita disposable income of rural households by region.* http://www.stats.gov.cn/tjsj/ndsj/2018/indexeh.htm. Accessed 29 May 2020.

Dowling, R. (2000). Power, subjectivity and ethics. In I. Hay (Ed.), *Qualitative research methods in human geography* (pp. 23–36). Oxford: Oxford University Press.

Evans, M. (1988). Participant observation: The researcher as a research tool. In J. Eyles & D. M. Smith (Eds.), *Qualitative methods in human geography* (pp. 118–135). Cambridge: Polity Press.

Goodman, J. (2010). *Grand Canyon of the East*. Kunming: Yunnan People's Publishing House.

Habich, S. (2016). *Dams, migration and authoritarianism in China: The local state in Yunnan*. London: Routledge.

Harwood, R. (2013). *China's new socialist countryside: Modernity arrives in the Nu River Valley*. Seattle: University of Washington Press.

He, X. (2007). New rural reconstruction and the Chinese path. *Chinese Sociology and Anthropology, 39*(4), 26–38.

Hennig, T., Wang, W., Magee, D., & He, D. (2016). Yunnan's fast-paced hydropower development: A powershed-based approach to critically assessing generation and consumption paradigms. *Water, 8*(476), 585–595.

Jacka, T. (2009). Cultivating citizens: Suzhi (quality) discourse in the PRC. *Positions: East Asia Cultures Critique, 17*(3), 523–535.

Jie, F., Zhenghai, H., Kerong, S., & Wei, S. (2013). Benefit-sharing mechanism of hydropower development: Nujiang Prefecture in Yunnan. *Journal of Resources and Ecology, 4*(4), 361–368.

Kearns, R. A. (2005). Knowing seeing? Undertaking observational research. In I. Hay (Ed.), *Qualitative research methods in human geography* (pp. 192–206). Oxford: Oxford University Press.

Kibler, K., & Tullos, D. (2013). Cumulative biophysical impact of small and large hydropower development in Nu River, China. *Water Resources Research, 49*(6), 3104–3118.

Klimpt, J.-E., Rivero, C., Puranen, H., & Koch, F. (2002). Recommendations for sustainable hydroelectric development. *Energy Policy, 30,* 1305–1312.

Kumar, A. (2018). Justice and politics in energy access for education, livelihoods and health: How socio-cultural processes mediate the winners and losers. *Energy Research and Social Science, 40,* 3–13.

Lim, K. F. (2014). Spatial egalitarianism as a social 'counter-movement': On socio-economic reforms in Chongqing. *Economy and Society, 43*(3), 455–493.

Liu, B., Liao, S., Cheng, C., Chen, F., & Li, W. (2018). Hydropower curtailment in Yunnan Province, southwestern China: Constraint analysis and suggestions. *Renewable Energy, 121,* 700–711.

Liu, K. (2004). *Globalization and cultural trends in China*. Honolulu: University of Hawai'i Press.

Luo, L., & Guo, W. (2013). Rural electrification in China: A policy and institutional analysis. *Renewable and Sustainable Energy Reviews, 23,* 320–329.

Mackerras, C. (2003). *China's ethnic minorities and globalization*. London: Routledge.
Malakar, Y. (2018). Evaluating the role of rural electrification in expanding people's capabilities in India. *Energy Policy, 114,* 492–498.
Oakes, T. (1998). *Tourism and modernity in China*. London: Routledge.
Ptak, T. (2014). Dams and development: understanding hydropower in far western Yunnan Province, China. *Focus on Geography, 57*(2), 43–53.
Ptak, T. (2019). Towards an ethnography of small hydropower in China: Rural electrification, socioeconomic development and furtive hydroscapes. *Energy Research and Social Science, 48,* 116–130.
Ptak, T., & Hommel, D. (2016). The trans-political nature of Southwest China's energy conduit Yunnan Province. *Geopolitics, 21*(3), 556–578.
Ptak, T., Yu, Y., & Su, X. (2019). Southwest China. In A. Orum (Ed.), *The Wiley-Blackwell encyclopedia of urban and regional studies*. Wiley-Blackwell: Malden.
Ritzer, G. (2011). *Globalization: The essentials*. Malden: Wiley-Blackwell.
Rousseau, J.-F. (2019). When land, water and green-grabbing cumulate: Hydropower expansion, livelihood resource reallocation and legitimisation in southwest China. *Asia Pacific Viewpoint, 61*(1), 134–146.
Statista. (2019). *Mobile phone penetration in China as share of the population from 2013 to 2019*. https://www.statista.com/statistics/233295/forecast-of-mobile-phone-user-penetration-in-china/. Accessed 29 May 2020.
Statistical Bureau of Yunnan Province. (2017). *Yunnan statistical yearbook*. Beijing: China Statistics Press.
Tilt, B. (2014). *Dams and development: The moral economy of water and power*. New York: Columbia University Press.
United Nations Education Scientific Cultural Organization (UNESCO) (2012). *Three Parallel Rivers of Yunnan World Heritage Areas*. https://whc.unesco.org/en/list/1083/. Accessed 29 May 2020.
Woods, M. (2007). Engaging the global countryside: Globalization, hybridity and the reconstitution of rural place. *Progress in Human Geography, 31*(4), 485–507.
Wu, F. (2015). *Planning for growth: Urban and regional planning in China*. London: Routledge.
Wu, F., & Phelps, N. (2011). (Post)suburban development and state entrepreneurialism in suburban Beijing's outer suburbs. *Environment and Planning A, 43*(2), 410–430.
Zhang, X., & Kumar, A. (2011). Evaluating renewable energy-based rural electrification program in western China: Emerging problems and possible scenarios. *Renewable and Sustainable Energy Reviews, 15*(1), 773–779.

Zhaohong, B., & Yanling, L. (2015). An overview of rural electrification in China: History, technology, and emerging trends. *IEEE Electrification Magazine, 3*(1), 36–47.

CHAPTER 9

As Time Goes by… Longitudinal Analysis of Dam Impacts Upon Livelihood Strategies in the Red River Valley

Jean-François Rousseau

Introduction

This chapter aims to reflect on and make sense of some medium-term impacts that hydropower development has brought upon ethnic minority villagers living along the shores of the Madushan Dam reservoir, built along the Red River in the Sino-Vietnamese borderlands. Like Tom Ptak (this volume), I posit that the overall consequences of hydropower development span beyond the sole and immediate impacts of this activity and the water and land access reshuffling it initially drives. Rather, I argue that hydropower development triggers and/or accelerates and/or intermeshes with wider livelihood and ontological reorganisations. Riparian societies therefore distinguish between 'before' and 'after' damming in discussions of a range of topics much broader than the dams themselves,

J.-F. Rousseau (✉)
University of Ottawa, Ottawa, ON, Canada
e-mail: jf.rousseau@uottawa.ca

J.-F. Rousseau and S. Habich-Sobiegalla (eds.), *The Political Economy of Hydropower in Southwest China and Beyond*, International Political Economy Series, https://doi.org/10.1007/978-3-030-59361-2_9

their most obvious consequences, and the impacts typically accounted for by the authorities.

As I unpack how this broader picture fluctuates over time, I find that hydropower expansion simultaneously contributes to strengthening and weakening certain aspects of the state modernist project of which it is a component. Riparian dwellers likewise maintain ambivalent feelings towards dams and the livelihood changes that they bring about. Dams are typically developed along with roads and spearhead a host of other development projects that villagers overall welcome—or at least come to consider as partly beneficial as time passes. Yet the dramatic water and land reallocations that dams drive make the impacts and livelihood disruptions of other development schemes, including transport infrastructure, pale in comparison with those of hydropower expansion (cf. Litzinger 2007). I unpack how interviewees' perceptions of these impacts emerged and evolved in six villages where reservoir creation had varying effects, with some villagers 'only' losing land and others forced to move to either self- or government-organised resettlement sites (see Habich-Sobiegalla and Rousseau, 2020).

Here I revisit ethnographic data acquired among riparian Handai communities from the Madushan reservoir area during nine ethnographic fieldwork journeys spanning 2009–2018. These individual visits to the Red River Valley lasted anywhere between a few weeks and three months and were distanced by no more than a year, except for a gap in 2012–2014. Each time, I was privileged to count on the help of a Handai research assistant whom I refer to as Li here.[1] With Li acting as a gate-opener, gate-keeper, and translator (from the local language to Mandarin Chinese), I was able to conduct hundreds of conversational interviews and a livelihood survey. I was also privileged to undertake participant observation and partake in daily activities with dozens of enthusiastic and welcoming research contributors—including the endless conversations, drinking, and eating involved in what Magnus Fiskesjö (2013) aptly coined participant intoxication (see also Turner 2013).

In this chapter, I detail how I foresee this contribution engaging with other researchers' efforts to take a step back and reflect on the broad picture that emanates from longitudinal data amassed through the years. I introduce the modernisation project currently unfolding in ethnic

[1] All names in this chapter are pseudonyms.

minority societies in the Sino-Vietnamese borderlands—with hydropower expansion as a key component. I then contextualise the research sites and key livelihood characteristics of the Handai dwellers that have contributed to my research. My results highlight how these individuals' experiences shape their perceptions of hydropower expansion and the development programme that dams partake in, and how these perceptions evolved in the decade following initial impacts. The analysis details how dams have consolidated components of the modernisation drive unfolding in the borderlands, and how they have weakened other dimensions of this government agenda.

Longitudinal Studies and Borderland Modernisation

Longitudinal studies are the exception rather than the norm in most scholarly literature. Revisiting and/or restudying the same fieldwork site(s) is notably costly, both in time and money (Rigg et al. 2012; Rigg and Salamanca 2015; Connell and Lee 2018; Brockington et al. 2019). Concentrating on a specific location might entail disincentives in terms of career development, especially for junior scholars aiming to secure tenure-track positions or tenure, as we are expected to 'move on' from our research after our PhDs. In some cases, tracking down research participants who have moved since earlier visits can be a challenge (Wilmsen and van Hulten 2017). Conversely, the long-term relations that revisiting scholars develop with some research contributors—including the one I have developed with Li—can also be regarded as creating bias. Yet, similar to other factors shaping a researcher's positionality (including gender, ethnicity, social status, etc.), I argue that such 'risk' can be mitigated through properly reflecting on how interpersonal relations shape qualitative researchers' subjectivities (Galdas 2017). Just as importantly, I find that long-term relationships between research contributors and researchers potentially yield greater opportunities for the latter to 'give back' and make research a more reciprocal endeavour (Swartz 2011).

Longitudinal ethnographic studies undertaken by anthropologists and human geographers are, relatively speaking, more frequent than other types of longitudinal research (cf. Rigg and Vandergeest 2012; Connell and Lee 2018).[2] Beyond the challenges that it entails, longitudinal research is acknowledged as a useful approach for probing how development yields differentiated and unanticipated results, how these results unfold and accumulate over time, and how individuals experience specific episodes as they occur (Murray Li 2014). As Rigg and Vandergeest (2012: 11–12) frame it:

> Restudies circumvent some of the pitfalls of extrapolating change from village studies, because they have more than one data point from which to view change. Rather than taking a snapshot in time and suggesting a trajectory of change by trying to predict the future, restudies take two or more points in the past to trace out an actual trajectory of transformation.

The notion of change these scholars highlight is central to many longitudinal studies probing how people and/or places evolve through time (Rigg and Vandergeest 2012; Connell and Lee 2018). The caveats associated with research collaborators' distant and potentially imprecise memories can thus be overcome, while the connections between personal histories and the political economies within which they occur stand out more clearly (Rigg and Salamanca 2015; Wilmsen 2016). No specific guidelines codify the length of engagement required for research to be considered as longitudinal; durations span from one to five decades in the literature cited above.

I personally investigate a decade-long interval spanning most of the 2010s. This is a period in which the Chinese state's modernisation drive within the Sino-Vietnamese borderlands has accelerated rapidly, with significant impacts on local ethnic minority societies. During the previous decade, the Go West Campaign (see this volume's introduction) set the conditions for this modernisation campaign. Go West channelled record amounts of investment into infrastructure, including transport and hydropower, in both the 'minority autonomous regions'[3] and 'remote border areas' that the campaign objectives specifically mention (Yao

[2] Although their authors do not emphasise this per se, the chapter by Brooke Wilmsen and her colleagues and that of Bettina Gransow in this volume also draw on longitudinal studies.

[3] The dams I discuss here are located in the Honghe Yi and Hani Autonomous Prefecture, which was granted this status given that significant portions of its population belong to the Yi and Hani ethnic minorities.

2009). Such investments overcame the 'friction of terrain' of the borderlands' mountainous geography—which has historically created obstacles to imperial and state projects alike—more thoroughly than ever before (see Scott 2009).

Unlocking the Sino-Vietnamese borderlands has served the state's ambition to better integrate the region's economy with the eastern seaboard and adjacent Southeast Asian states, including through hydroelectricity exports (see Summers 2019; Hennig and Magee, this volume). In parallel, discursive and financial incentives promote a socio-cultural model that emulates the Han Chinese, on the grounds that 'modern' lifestyles are more appropriate for the capitalist economy than customary 'backward' practices. In these borderlands' ethnic minority settlements, prevalent modern/hybrid crop varieties, minivan and electric appliance ownership, Wi-Fi, and even beer drinking all testify to this model gaining traction (see Turner et al. 2015; Ptak, this volume). These count as 'privileges' made possible through what Yeh (2013) calls the 'gift of development' that the 'benevolent' Chinese state has provided.

Yet the gift comes with strings attached, such as the expectation that ethnic minorities show gratitude for being invited to contribute to national development objectives, including providing distant urban areas with a 'clean' and reliable energy source (see Yeh and Wharton 2016; Habich-Sobiegalla and Rousseau, 2020). In this specific case, gratitude notably entails not disturbing social stability and internalising the myth that Southwest China's 'abundant' natural resources are best used by and for external actors rather than local ones. As damming occurred along the Red River, the Handai people I now turn to experienced a series of livelihood reorganisations that unfolded in unexpected ways throughout the 2010s.

Handai People and Red River Dams

While most individuals presented in the earlier chapters of this book section have customarily lived in the Yunnan highlands (see chapters by Harlan and Ptak, this volume), the Handai settlements I focus on are in lowland areas along the narrow Red River floodplain.[4] Although

[4] Globally, state-led social hierarchisations have long instrumentalised the highland-lowland dichotomy, with determinist attitudes continuing regarding highlanders' 'wildness' and lowlanders' 'civilisation' (Michaud et al. 2016). These rigid statements do not account

long isolated from urban settlements, this setting has historically been, relatively speaking, easier to access than adjacent higher grounds. Trade caravans travelling along the southern Silk Road passed through the area in precolonial times (Yang 2009). Later, relative accessibility also meant that some technologies reached the lowlands earlier than the highlands (Henin and Flaherty 1994; Yang 2009). For instance, small hydropower dams were first built in this area as early as in the late 1950s, and electricity has first been available in the villages about a decade later. Hybrid seeds first appeared in the valley during the economic reform era. Peasants' reactions to these crops were then lukewarm, mainly because they had no knowledge on how to grow them and felt that there was no way they could make them profitable in the capitalist economy to which they were now exposed.

Living in the lowlands and along the Red River floodplain also meant that Handai farmers customarily enjoyed different agricultural conditions from societies established just a few hundred metres higher in the nearby mountains. This is because the lowland climate is warmer, the growing season is longer, land is flatter, and the Red River's flood and pulse regime used to bring sediment to rejuvenate soil yearly. Hence, whereas swidden farming centred on maize was long the primary farm system in the highlands, Handai could grow wet rice on the same land yearly. Handai research collaborators convey that this led to Handai people being more attached to specific patches of land and less prone to engaging in work migration than neighbouring ethnic groups (see Michaud 2011). Handai also customarily shared a common agrarian calendar centred on wet rice farming, which involves intensive work periods during which labour exchanges are common and forge social relationships and social capital in specific ways (Bray 1986; Talhelm et al. 2014).

Although they self-identify as such, Handai people are not officially acknowledged within the Chinese state ethnic minority classification system, partly because they consist of a very small and geographically localised population. They are thus subsumed within the broader Dai 'minority nationality' (*shaosu minzu*), most likely due to their language

for the plethora of interactions and movements between societies customarily settled in both biotypes. I refer to this categorisation here only to highlight the geographical differences between the Red River Valley and the settings where the other chapters of this section are set, and how these circumstances shape ethnic minorities' encounters with hydropower expansion.

being akin to those of other Dai subgroups. Handai comprise the most important ethnic group along the 60 km-long Madushan reservoir where this research is set. The Madushan Dam (288 MW) was completed in 2011, two years after the upstream 150 MW Nansha Dam became the first hydropower dam built along the mainstream of the Red River. Both projects have driven environmental changes impacting Handai livelihoods in different ways, although riparian dwellers only received direct compensation for the flooding of their land and/or houses (see Rousseau 2020). I now probe in more detail how some of these impacts unfolded and how people reacted through time.

Dam Impacts Through Time

In looking back and re-analysing the data I gathered in the Madushan reservoir from 2009 to 2018, I distinguish between how people foresaw the dams and their impacts in the initial years after damming began (2009–2012) and half a decade later (2014–2018). The data I introduce below are organised accordingly.

The First Years

The time interval between the completion of the upstream Nansha Dam and the creation of the Madushan reservoir was perceived as a good period for the Handai villagers who would not need to relocate due to hydropower expansion. Financial compensation for land areas to be flooded by the future Madushan reservoir were distributed upfront, providing households with a lump sum equalling many times their yearly average income. This massive cash influx triggered a significant house renovation spree in many villages, with people eager to replace their old clay-brick houses with 'modern' ones made of concrete. Meanwhile, they could still tend their crops as before, and the filling of the upstream Nansha reservoir even allowed them to access more land, as the Red River's discharge was maintained at lower levels than usual.

Things were more complicated for those who lived in villages located in to-be flooded areas; most of those who needed to resettle chose to migrate to a nearby resettlement built by government-hired contractors.[5]

[5] For a discussion of households who chose not to move out and to self-resettle, see Habich and Rousseau (2020).

As had happened earlier elsewhere in the Red River Valley and beyond, resettlers were awarded sub-par land and houses (Habich 2016). Yet isolated protests, including attempts to not move on the date set by the local authorities, were rapidly quelled when armed police were mobilised. As one villager who did not want to leave at first remembers, "this is when we understood that we had no more choice" (Interview, 16 August 2017). While they needed to replant their new landholdings with banana, the main cash crop in the area, those resettling to the new village could at least continue to cultivate their previous land until flooding occurred.

Then flooding came, affecting each village and villager differently, with some places flooded entirely and others only partly. Yet almost everyone was impacted, and consequences were aggravated by the steep slopes of the Red River Valley limiting possibilities to access new land. At the onset of the 1980s economic reforms, households had been awarded 0.8 *mu* of land per household member. However, as agricultural intensification and land parcelling proceeded apace in the following decades, average per capita landholdings averaged only 0.5 *mu* in three villages I surveyed in the early 2010s. Thus, land scarcity had already emerged as a principal concern before the Madushan reservoir flooded half of the landholdings available in the research villages. In a context where land is considered the single most important household asset due to both its economic importance and the value people attach to it, this was a traumatic episode.

Villagers were compelled to reorganise their livelihoods to compensate for this loss. Most were confronted with the dilemma between staying home or engaging in work migration. Close connection to their land, combined with higher agrarian incomes in the valley compared to higher elevations, had until then created few incentives for people to move out. Those villagers who resorted to moving to nearby or distant cities to find low-skilled jobs were almost exclusively those who faced the most difficult financial situations, including people who had lost all their landholdings to the reservoir. Those who stayed acknowledged that they were privileged to not have had to emigrate, as those who left had no other choice.

The most frequently encountered livelihood diversification strategies among villagers who remained in the valley were centred on agriculture and reservoir fisheries, with most villagers adopting one or more of the following tactics: opening new land in sloped and poorly irrigated areas; investing in fish farming to take advantage of the newly created reservoir hydrology that now made this activity possible; engaging in reservoir fisheries; or further intensifying their agricultural activities.

All of these individual strategies came with drawbacks. First, in the context of the ongoing drought of the early 2010s, cultivating the poor and insufficiently irrigated soil of the sloping lands turned out to be impossible despite the significant hard work and financial investments that locals sank into developing irrigation networks. Second, aquaculture in the reservoir also turned out to be a financially draining and labour-heavy activity that yielded few positive outcomes. Irregular chemical releases from an upstream sugar plant randomly killed farmed fish populations, discouraging those who had initially mobilised the resources needed to pursue this activity. Third, reservoir fisheries also suffered from the same pollution problem, though to a lesser extent. Nonetheless, as a boom in fishing occurred, the population of Red River catfish—the area's most valuable endemic fish[6]—declined sharply. Both factors combined to render fishing a volatile, non-lucrative livelihood.

Agricultural intensification entailed a shift from growing food to cash crops, with hybrid banana varieties emerging as the most sown species and wet rice being widely abandoned. The quantities of chemical inputs that people sprayed on their land also rose significantly after flooding occurred. This was also when hand tractors replaced water buffalo as the main tilling technology. Riparian farmers viewed these changes ambivalently. On the one hand, people had reservations about the quality of the food they now predominantly produced, preferring land-race rice, fruit, and vegetable varieties to those that they now sold and bought in much greater quantities than before for personal consumption. The negative impacts of using chemical fertilisers also created concerns. Further, the need to buy farm inputs and a greater share of the food they ate increased their exposure to the vagaries of food and oil markets. On the other hand, greater agricultural commercialisation allowed some villagers to compensate for income loss associated with land flooding. Also, people felt that the use of 'modern' farming techniques, the reduced planting of rice, and the rise in smaller landholdings meant that villagers were less busy working in the fields than in the past.

In parallel, road infrastructure development steadily reduced travel times between the provincial capital of Kunming and Honghe Prefecture, including the UNESCO World Heritage Yuanyang rice terraces in

[6] The Red River catfish is a migratory species akin to the more widely-known Mekong catfish, whose migration routes are also threatened by hydropower development.

the highlands adjacent to the Red River Valley. All these circumstances combined in shaping further livelihood changes as the years went by.

The New Normal

The main livelihood strategies deployed in the initial years following the construction of the Red River dams were centred on fishing and agrarian activities. Skyrocketing reservoir fish productivity sustained a short-lived fishery boom that went bust within just a few years, replicating a scenario typical in newly created reservoirs (Scudder 2005).[7] Also, whereas most households maintained agricultural activities, these decreased in importance as local livelihood portfolios increasingly shifted toward non-agricultural activities.

The most significant change beyond this trend was a drastic rise in the number of people engaging in work migration from the mid-2010s onwards. This change came along with new villager perceptions towards leaving the village to find low-skilled work in the service, construction, and manufacturing sectors in cities in Yunnan and beyond. Earlier attitudes were that "people only leave the village if they cannot make a living here anymore" (Interview, 17 March 2012). Yet, in the course of only a few years, migration emerged as a reasonable solution for reallocating household resources in the light of the sharp decrease in farm labour intensity highlighted above. New discourses about farming also emerged. Whereas those that kept access to some land after reservoir creation reported that farming still allowed them to make a decent livelihood in the village after flooding, this had changed a few years later. Notably, this was because people were spending more money on basic necessities such as food, education, and health care. However, they were also spending more on buying, operating, and maintaining higher-end commercial goods rapidly diffusing in the villagescapes, including solar water heaters, air conditioners, at-home Wi-Fi networks, small electric or gas-powered three-wheel tractors, and motorcycles.

Discussing migration, Li (the research assistant working with me), who himself had taken up a job in an electronic component factory in Suzhou (Jiangsu), explained in 2016: "Lots of people have gone away these last few years. The young generation is not interested in working in the

[7] This is because flooding at first allows for nutrient release and rising fish populations.

fields anymore" (Interview, 7 November 2016). As this shift happened, migrants built on social networks to find job opportunities in the same businesses as those employing their relatives and co-villagers, and sought shared housing with villager migrants who had arrived earlier. With no exception, the migrant villagers I interviewed considered work migration a difficult journey that involved being away from their relatives, very long days, shoddy housing conditions, and low pay. All considered this a transient strategy undertaken for some time to amass enough money to ultimately invest in new village-based income-generating assets. Critiques about migration being inappropriate for Handai people were replaced by arguments framing this strategy as a solution to achieving a steady financial income in the short and long terms. Migrant workers whose children stayed in the villages would send some of their income back to their relatives regularly. In contrast, younger migrants reported that leaving the village and enjoying financial autonomy allowed them to discover urban life.

Fuelled by steadily growing road traffic, retail was another non-agricultural sector that developed swiftly in the villages. Unimpeded road enhancement throughout Honghe Prefecture has allowed for further tourism development in the nearby highlands and rapid population growth in the nearest urban settlement, the county-level city of Nansha, located 10–25 km away from the research villages. A minority of villagers had long sold predominantly local fruits and delicacies to passers-by in rudimentary stalls along the main road going through the villages. In the early 2010s, about a third of the people I surveyed reported that at least one member of their household sold mostly self-produced foodstuffs along the road at their convenience. By 2016, in the villages with the best locations along the road, at least one member of every household was involved in full-time daily fruit trading. In order to ensure a steady supply of fruit throughout the year, most of the products traded now came from afar, including Vietnamese mango, banana, and tamarind that "look and sell better than those we produce here, although they don't taste as good" according to one person involved in such trading (Interview, 15 May 2018).

By 2015, fruit stalls installed by petty traders along the road to protect them from rain and sun had become notable structures in the local villagescapes. Local authorities, however, removed them as part of the 'beautiful homeland' (*meili jiayuan jiashe*) village renovation schemes then being implemented throughout the Madushan reservoir area. As

Habich-Sobiegalla and Plümmer (this volume) explain, authorities saw participation in this campaign as a privilege that came with responsibilities. In this case, responsibilities included removing the stalls, and the privilege granted was their replacement with a permanent 'fruit market' adorned with the bright colours and symbols of Dai ethnic culture that authorities had decided upon for the area.[8] It was argued that the old stalls were 'messy' (*luan*), projected a bad image of the village, and affected traffic fluidity and road safety. The villagers strongly disagreed with these assessments and argued instead that the authorities had begun paying attention to the stalls only because trade had grown rapidly and they were eager to claim part of the profits for themselves. This explains, they argued, why a monthly tax was now being levied on traders.

By 2017, the local village retail sector remained important, though traders that could not secure a good spot in the new market's first alley—the one closest to the road—argued that their business had gone down. They explain that many customers do not get out of their cars and instead hail the first trader they see alongside the road. By 2018, fences had been placed between the market and the road to discourage this behaviour, considered risky. However, traders located in inner alleys still reported that their business was lower than in the past. In the same year, fruit sellers had also started to raise concerns about fruit trading. Construction had just begun on a new highway project linking the two closest district-level towns of Gejiu and Luchun, and traders feared that transit traffic would plunge drastically upon the project's completion in 2021. They foresaw many people having to reorganise their livelihoods yet again. At that stage, no one could figure out what village-based activity might allow them to cope. But all laughed at my inquiries about whether agriculture could be a solution.

Connecting the Dots

A key tenet of the scholarship on water's role in nature-society relations claims that water is a political field; power relations oversee water allocation together with the benefits and consequences emerging from changes in water access regimes (Loftus 2009; Ballestero 2019). Powerful actors are prone to (re-)allocating water to economically productive

[8] As it turned out, peacocks are one of the main symbols of 'Dai-ness' that the state opted for, although they are absent from local culture (see Rousseau and Turner 2018).

and state-making sectors through large-scale technical interventions such as hydropower dam construction. In doing so, they emphasise specific knowledge regimes and visions of the hydraulic, environmental, and livelihood dimensions of such endeavours (Ballestero 2019). Yet the resulting discourses consist of simplifications that overlook the complex interconnections between water and the human and non-human worlds (Bakker 2012). In some cases, simplifications emerge from a goal-oriented strategy aimed at ensuring that hydraulic projects proceed apace and at a low cost, regardless of their socio-environmental impacts. For instance, the environmental impact assessments (EIA) drafted prior to dam construction—at best, and a posteriori at worst—sometimes consist of no more than caricatural summaries of the consequences of these projects (see Tilt 2014; Rousseau 2020). Even when EIAs consist of more than a cursory, politically-driven presentation of data, it remains a challenge to anticipate the interconnected impacts of hydrological changes before they occur. Longitudinal analysis does not bring such anticipatory power, but it can yield important insights that help comprehend and interpret how impacts manifest over time.

Building from this approach, the data introduced above probe livelihood changes that have occurred in the wake of hydropower dam construction along the Red River. These changes of course did not happen in a decade-long vacuum where hydropower development was the sole driver of livelihood change (see Wilmsen 2018). Rather, changes unfolded in a specific political economy characterised, notably, by the state's modern and capitalist projects rapidly gaining ground within ethnic minority societies in the Sino-Vietnamese borderlands. Hydropower dam construction and reservoir creation partake in these projects. Regardless of such 'chicken or egg' causation questions, hydropower expansion clearly had profound and specific influences on the livelihood trajectories that Handai research collaborators adopted as they experienced the combined influences of various 'modern' state projects. The specific socio-environmental context in which this occurred adds to the circumstances that shaped riparian dwellers' reactions and their evolution throughout the 2010s.

Two overarching trends emerge from the longitudinal data above: first, riparian dwellers have undergone livelihood diversification strategies that have led to non-agrarian activities gaining importance in their livelihoods; second, Handai livelihoods now build from a wider array of activities and assets. These conclusions echo scholarship on livelihood de-agrarianisation and pluralisation (Ellis 2000; Rigg 2001), and, more broadly, rural livelihood and peasant studies scholarship conceptualising how peasants experience and engage with various stages and models of capitalist development and associated (neo-)liberal governmentalities (Borras 2009; Bennike et al. 2020). As the years went by after the creation of the Madushan reservoir, these two pathways correlated with the emergence of a third trend, namely the spatial spreading of Handai livelihood strategies.

Investigating how these pathways evolved over a decade allows us to further reflect on their causes and implications further. As my data highlight, initially de-agrarianisation, pluri-activity, and migration were not the Handai's preferred strategies for addressing the impacts of reservoir-induced land flooding. Confident that the local environment could allow them to cope and maintain their livelihoods despite their decreased landholdings, those that still maintained access to land had a natural inclination towards turning to this asset to cope. This reflects the particular relationship between the villagers and their landholdings, which, beyond their dusty materiality, also consist of assets that farmers resort to in times of hardship (see Tilt 2010; Hall et al. 2011). Likewise, the fact that villagers hung on to their remaining landholdings—with virtually no contract land transfers occurring in spite of many people desperately searching for new ways to gain additional income—testifies to the importance that land and its intergenerational transfer hold for these individuals.

The quest for new lands, expansion of fish farms, establishment of reservoir fisheries, and agricultural intensification that occurred in and around the newly created reservoir all point to a single pathway. Indeed, these strategies highlight how villagers acknowledged that maintaining customary farming practices on their reduced landholdings would no longer suffice to maintain their livelihoods. Yet these decisions also testify to villagers' beliefs that producing foodstuffs for market purposes would be economically viable pending sufficient production and/or productivity increases. The resources that villagers expended in trialling these solutions, together with the trade-offs they committed to when further intensifying their agricultural practices, all testify to their early aspirations to maintain agrarian livelihoods. Yet their limited asset base and the difficulties encountered when trying to expand it ultimately made this impossible and convinced them that they needed to further reorganise their livelihoods.

In the meantime, villagers became much more involved with the market economy than ever before. This led to a reduction in the farm-related workload, while requiring that villagers secure greater cash incomes to maintain their livelihoods. Both aspects combined in driving significant change in the perceptions that villagers held towards work migration. Considered a testimony to someone's incapacity to make ends meet at home in the early 2000s,[9] migration became a popular diversification strategy just a few years later. People felt that migration entailed reasonable trade-offs between being away from the village and enduring difficult work and living conditions on the one hand, while securing a stable income on the other hand. The development of market-based retail activities also testifies to the same need to secure increased income, although this village-based solution is somewhat more aligned with earlier preferences for on-site livelihood strategies.

Just as growing market integration made agricultural livelihoods increasingly redundant, agricultural intensification, increasing out-migration, and the growing importance of village-based retail activities were part of a thorough reorganisation of social capital and networks. Previously, the customary farm system centred on wet rice had created strong social binders because all villagers shared the same agricultural calendar with communal light and heavy periods of labour. This orientation away from traditional labour exchange mirrors Tilt and Gerkey's (2016) research along the Lancang cascade, where dam building led to the erosion of this asset.

Though the switch to cash crops did lead to a significant reduction in farm labour exchanges along the Madushan reservoir, this has not heralded the complete dismantling of social capital beyond the household or extended household scales. My data instead point to the multiplication and geographical extension of clusters grouping individuals who share daily activities and resources. These clusters now span wider areas, encompassing the reservoir as well as the cities in Yunnan and beyond where villagers now work. In addition, these different clusters are potentially more exclusive than in the past, since as a migrant interviewee explained when she came back home during the spring festival, "village life has

[9]This perception then contrasted with what prevailed in most of China, as reforms to the *hukou* household registration begun in the late 1980s had led, by the late 2000s, to 150 million rural citizens engaging in work migration to urban areas (Chan 2009).

changed, now people either work elsewhere or the other women are too busy selling at the market all day" (Interview, 18 February 2015).

This quote highlights how people have reservations regarding the changes that they have experienced during the 2010s and, ultimately, regarding the modernisation project shaping these. Regardless of their frequent reorganisation of livelihood activities, none of their tactics have yet allowed them to achieve a livelihood that they could foresee maintaining for a long duration. Farm-based strategies and reservoir fisheries have failed, work migration is considered a transient pathway, and the village-based retail sector faces a bleak forecast.

Conclusion

Reflecting on what developments could help villagers make a living over the long term, Li foresaw that tourism was probably the only 'sustainable' option. He felt that given the forthcoming highway and drop in transit traffic, and with farming as well as fish production hitting a ceiling, the villages needed to attract overnight tourists rather than just passers-by. From an external point of view, this makes sense given that hydroelectric reservoirs are often the locus of tourism development in China and beyond.

Yet Li felt that this tourism pivot would not materialise. He argued that the prefecture government had no plan to emphasise valley tourism and encourage tourists to visit the Yuanyang terraces. He blamed this on tensions between local cadres and villagers since the impacts from the reservoir first arose, together with recurrent villager critiques of local governance. In other words, the villagers' lack of 'gratitude' tested the benevolence of the authorities and eroded their desire to promote village-based, long-term livelihood options.

Relatedly, villagers argue that poor long-term project coordination has sometimes created burdens for them. For instance, *meili jiayuan* involved that all houses in the villages be repainted at the government's expense. Yet most houses had been renovated and painted less than a decade earlier at villagers' expense. In retrospect, many argued that they could have saved significant amounts of money had they known about the government's project earlier. Here, the villagers build on memories to reach ambivalent conclusions toward the project.

The above examples further confirm that the ways in which authorities plan and implement development schemes, as well as the alignment of these schemes with reservoir populations' livelihood priorities and strategies, are anchored in a series of events that occur over a long duration. I argue that longitudinal analysis is one approach that researchers could mobilise further to better appreciate how these circumstances manifest and evolve. It is my hope that this paper will encourage researchers with longitudinal datasets to go back and revisit their findings in the light of circumstances different than those that prevailed when they first filled up their old notebooks.

References

Bakker, K. (2012). Water: Political, biopolitical, material. *Social Studies of Science, 42*(4), 616–623.

Ballestero, A. (2019). The anthropology of water. *Annual Review of Anthropology, 48*(1), 405–421.

Bennike, R. B., Rasmussen, M. B., & Nielsen, K. B. (2020). Agrarian crossroads: Rural aspirations and capitalist transformation. *Canadian Journal of Development Studies/Revue canadienne d'études du développement, 41*(1), 40–56.

Borras, S. M. (2009). Agrarian change and peasant studies: Changes, continuities and challenges—An introduction. *Journal of Peasant Studies, 36*(1), 5–31.

Bray, F. (1986). *The rice economies: Technology and development in Asian societies*. Oxford: Blackwell.

Brockington, D., Coast, E., Mdee, A., Howland, O., & Randall, S. (2019). Assets and domestic units: Methodological challenges for longitudinal studies of poverty dynamics. *The Journal of Peasant Studies*, 1–21.

Chan, K. W. (2009). The Chinese hukou system at 50. *Eurasian Geography and Economics, 50*(2), 197–221.

Connell, J., & Lee, H. (Eds.). (2018). *Change and continuity in the Pacific: Revisiting the region*. London: Routledge.

Ellis, F. (2000). *Rural livelihoods and diversity in developing countries*. Oxford: Oxford University Press.

Fiskesjö, M. (2013). Gifts and debts: the morality of fieldwork in the Wa lands on the China-Burma Frontier. In S. Turner (Ed.), *Red stamps and gold stars: Fieldwork dilemmas in upland socialist Asia* (pp. 61–79). Vancouver: UBC Press.

Galdas, P. (2017). Revisiting bias in qualitative research: Reflections on its relationship with funding and impact. *International Journal of Qualitative Methods, 16*, 1–2.

Habich, S. (2016). *Dams, migration and authoritarianism in China: The local state in Yunnan*. New York: Routledge.

Habich-Sobiegalla, S., & Rousseau, J.-F. (2020). Responsibility to choose: Governmentality in China's participatory dam resettlement processes. *World Development, 135*, 1–11.

Hall, D., Hirsch, P., & Murray Li, T. (2011). *Powers of exclusion: Land dilemmas in Southeast Asia*. Singapore: NUS Press.

Henin, B., & Flaherty, M. (1994). Ethnicity, culture, and natural resource use: Forces of change on Dai society, Xishuangbanna, Southwest China. *Journal of Developing Societies, 10*(2), 219–235.

Litzinger, R. (2007). In search of the grassroots: Hydroelectric politics in Northwest Yunnan. In E. J. Perry & M. Goldman (Eds.), *Grassroots political reform in contemporary China* (pp. 282–299). Cambridge: Harvard University Press.

Loftus, A. (2009). Rethinking political ecologies of water. *Third World Quarterly, 30*(5), 953–968.

Michaud, J. (2011). Conclusion: Lesson for the future. In J. Michaud & T. Forsyth (Eds.), *Moving mountains: Ethnicity and livelihoods in highland China, Vietnam, and Laos* (pp. 215–237). Vancouver: UBC Press.

Michaud, J., Swain, M. B., & Barkataki-Ruscheweyh, M. (2016). *Historical dictionary of the peoples of the Southeast Asian massif*. Lanham: Rowman & Littlefield.

Murray Li, T. (2014). *Land's end: Capitalist relations on an indigenous frontier*. Durham: Duke University Press.

Rigg, J. (2001). *More than the soil: Rural change in Southeast Asia*. Harlow: Prentice Hall.

Rigg, J., & Salamanca, A. (2015). The devil in the detail: Interpreting livelihood turbulence from a 25-year panel study from Thailand. *Area, 47*(3), 296–304.

Rigg, J., Salamanca, A., & Parnwell, M. (2012). Joining the dots of agrarian change in Asia: A 25 year view from Thailand. *World Development, 40*(7), 1469–1481.

Rigg, J., & Vandergeest, P. (Eds.). (2012). *Revisiting rural places. Pathways to poverty reduction and prosperity in Southeast Asia*. Singapore: NUS Press.

Rousseau, J.-F. (2020). When land, water and green-grabbing cumulate: Hydropower expansion, livelihood resource reallocation and legitimisation in southwest China. *Asia Pacific Viewpoint, 61*(1), 134–146.

Rousseau, J.-F., & Turner, S. (2018). Not at all costs: Frontier modernization schemes and ethnic minority livelihood debates in the Sino-Vietnamese borderlands. *Verge: Studies in Global Asias, 4*(1), 133–159.

Scott, J. C. (2009). *The art of not being governed: An anarchist history of upland Southeast Asia*. New Haven: Yale University Press.

Scudder, T. (2005). *The future of large dams: Dealing with social, environmental, institutional and political costs*. London: Earthscan.

Summers, T. (2019). The belt and road initiative in Southwest China: Responses from Yunnan province. *The Pacific Review*, 1–24.

Swartz, S. (2011). 'Going deep' and 'giving back': Strategies for exceeding ethical expectations when researching amongst vulnerable youth. *Qualitative Research, 11*(1), 47–68.

Talhelm, T., Zhang, X., Oishi, S., Shimin, C., Duan, D., Lan, X., et al. (2014). Large-scale psychological differences within China explained by rice versus wheat agriculture. *Science, 344*(6184), 603–608.

Tilt, B. (2010). *The struggle for sustainability in rural China: Environmental values and civil society*. New York: Columbia University Press.

Tilt, B. (2014). *Dams and development in China: The moral economy of water and power*. New York: Columbia University Press.

Tilt, B., & Gerkey, D. (2016). Dams and population displacement on China's Upper Mekong River: Implications for social capital and social-ecological resilience. *Global Environmental Change, 36*, 153–162.

Turner, S. (2013). *Red stamps and gold stars. Fieldwork dilemmas in upland socialist Asia*. Vancouver: UBC Press.

Turner, S., Bonnin, C., & Michaud, J. (2015). *Frontier livelihoods: Hmong in the Sino-Vietnamese borderlands*. Seattle: University of Washington Press.

Wilmsen, B. (2016). After the deluge: A longitudinal study of resettlement at the Three Gorges Dam, China. *World Development, 84*, 41–54.

Wilmsen, B. (2018). Is land-based resettlement still appropriate for rural people in China? A longitudinal study of displacement at the Three Gorges Dam. *Development and Change, 49*(1), 170–198.

Wilmsen, B., & van Hulten, A. (2017). Following resettled people over time: The value of longitudinal data collection for understanding the livelihood impacts of the Three Gorges Dam, China. *Impact Assessment and Project Appraisal, 35*(1), 94–105.

Yang, B. (2009). *Between winds and clouds: The making of Yunnan (second century BCE to twentieth century CE)*. New York: Columbia University Press.

Yao, Y. (2009). The political economy of government policies toward regional inequality in China. In Y. Huang & A. M. Bocchi (Eds.), *Reshaping economic geography in East Asia* (pp. 218–240). Washington, DC: World Bank.

Yeh, E. T. (2013). *Taming Tibet: Landscape transformation and the gift of Chinese development*. Ithaca: Cornell University Press.

Yeh, E. T., & Wharton, E. (2016). Going west and going out: Discourses, migrants, and models in Chinese development. *Eurasian Geography and Economics, 57*(3), 286–315.

PART III

Transnational Matters

CHAPTER 10

Technical and Policy Constraints on the Role of Chinese Hydropower in a Renewable Mekong Region

Darrin Magee

Introduction

Dams in China and Southeast Asia, as elsewhere in the world, have a mixed legacy. In some cases, dams have provided important infrastructure for storing irrigation water, controlling flooding, or producing electricity in remote areas where access to national grids has historically been limited or non-existent (Magee 2014; see also Ptak, this volume). Yet they have also displaced human communities, disrupted aquatic and riparian habitats, altered the flow of water and sediment, and brought numerous other negative socio-economic, biophysical, and geopolitical consequences. These impacts have long been recognised in the scholarly literature outside China (Goldsmith and Hildyard 1986; McCully 1996; World Commission on Dams 2000; Scudder 2005; Brown et al.

D. Magee (✉)
Hobart and William Smith Colleges, Geneva, NY, USA
e-mail: MAGEE@hws.edu

J.-F. Rousseau and S. Habich-Sobiegalla (eds.), *The Political Economy of Hydropower in Southwest China and Beyond*, International Political Economy Series, https://doi.org/10.1007/978-3-030-59361-2_10

2009; Kibler and Tullos 2013), and are increasingly visible in the Chinese scholarly literature as well (see, for example, Shuai 2017). Due to their complexity and long timeframe from initial planning to construction to commissioning, dams may also be sites of corruption, where unscrupulous officials divert funds that would otherwise be destined for vital functions such as compensating resettled villagers or ensuring safe construction practices.

Dams are built to last; their impacts, both positive and negative, transcend time and space. Many of the world's largest dams were built nearly a century ago and are now showing their age (Poff and Hart 2002; Ho et al. 2017). The designers of Three Gorges Dam, the largest dam in the world by many measures, boasted that it would control the mighty Yangtze for centuries, even though faster-than-expected siltation rates now make such estimates seem overly optimistic. When dams are built on transboundary rivers such as the Mekong, Salween, Ayeyarwady (Irrawaddy), and Brahmaputra, negative impacts may be felt hundreds of kilometres away in different countries whose governments and citizens may have little to no recourse against the upstream country. Such is clearly the case in the Greater Mekong Subregion (GMS),[1] where downstream countries Myanmar, Laos, Thailand, Cambodia, and Vietnam largely perceive dam construction upstream in as a unilateral activity by Chinese dam-builders who minimise or ignore legitimate concerns of downstream actors. In almost all cases, except perhaps micro-hydropower dams and small weirs for water storage, negative socio-economic and biophysical impacts are most acutely felt in close proximity to the dam and its impoundment, whereas positive benefits such as downstream flood control and improved electrical grid reliability are more diffuse and accrue to a broader, more geographically dispersed population.

As Sebastian Biba (this volume) discusses, the governance of dams throughout China and the Greater Mekong Subregion has been anything but comprehensive. Small and medium-sized dams on tributaries are often approved at local levels of government with little to no attention paid to their impacts on downstream communities, hydrology, or ecology. Research has demonstrated clearly, though, that the cumulative

[1] In this chapter, I use the terms GMS, Lancang Mekong River basin (LMRB), and Mekong region essentially interchangeably as a matter of geographic convenience. In reality, GMS is a discursive creation of the Asian Development Bank in 1992 and predates China's LMRB framework and subsequent LMRB Cooperation by nearly two decades.

impacts of multiple small dams, especially in terms of habitat fragmentation, may far exceed the impact of one large dam (Kibler and Tullos 2013). Cumulative impacts of dams large and small in the same watershed remain poorly understood, but the research that does exist on the topic tends to paint a bleak picture for riparian ecosystems. Institutional frameworks in the Greater Mekong Subregion that might facilitate co-management and forward-looking governance are hampered by a history of mistrust in the region, especially between the downstream countries (Vietnam, Cambodia, Laos, and Thailand) and China, with Myanmar caught somewhere in the middle.[2] Perhaps nowhere is this distrust, and its implications for governance, more clearly evidenced than in the Mekong River Commission, which counts as members only the four downstream countries while China and Myanmar participate nominally as observers.

Yet dams are, as noted above, here to stay. Indeed, the pressure to continue building more large hydroelectric dams in southwestern China and elsewhere in the region and around the world will almost certainly continue to grow for two reasons. First, Chinese dam developers have, since the turn of the twenty-first century, been strongly encouraged by central policy to go westward and outward. Their race to develop the rivers of southwestern China goes hand-in-hand with the development of electricity-intensive mining and smelting operations in mineral-rich Yunnan (Hennig et al. 2013; see also Hennig and Magee, this volume). Second, dams are particularly well suited to complementing wind and solar power, whose variability and intermittency can hamper their large-scale integration into existing power grids. Thus, as the costs of wind and solar continue to decline and the number of utility-scale wind and solar farms continues to grow, the need will grow proportionately for generators that can smooth their intermittency and make wind and solar more attractive to wholesale and retail purchasers of renewable electricity.

This chapter examines some of the technical and policy roadblocks that have hampered effective utilisation of southwestern China's hydropower

[2] The halting of work on the controversial Myitsone Dam, a 6000 MW project planned by China Power Investment for the Irrawaddy River in Myanmar's Kachin State, provides insight into the complexities of Chinese influence on dam projects *vis-à-vis* downstream countries. The project, which International Rivers (n.d.: online) called "a second Three Gorges project for China – this time in Burma" was suspended, much to the consternation of the Chinese developers, after a public outcry in Myanmar and abroad about the damage it would do to the Irrawaddy ecosystem.

assets, which are currently overbuilt and under-utilised (Magee and Hennig 2017). In section "Electricity Generation in Southwestern China and Mainland Southeast Asia", I give an overview of electricity generation and consumption in China's Southwest and throughout the Greater Mekong Subregion. In section "Balancing Variable Renewables with Hydropower", I discuss the technical and policy mechanics of balancing intermittent renewables like wind and solar with hydropower. In sections "Hydropower's Role in Enabling Wind and Solar Power in China" and "Going Outward: Using Yunnan Hydropower to Enable a Solar Transition in Cambodia", I explore these general considerations in the specific geographic contexts of China, whose wind and solar installations have outpaced those of the rest of the world in recent years, and downstream Cambodia, whose significant untapped solar potential presents an alternative to the dam-centred development model pushed by Chinese power companies in the Mekong basin. Finally, I conclude by summarising recent developments in southwestern China's hydropower sector and the country's power sector at large, along with their implications for greater reliance on renewable energy in the rapidly developing Mekong region.

Electricity Generation in Southwestern China and Mainland Southeast Asia

The generation of electricity across southwestern China and Mainland Southeast Asia varies, not surprisingly, according to geography, level of economic development, availability of various energy resources, and the sophistication of generation, transmission, and distribution assets ('grid'). Figure 10.1 shows electricity generation by country for the five countries of the Mekong basin downstream of China. China is excluded from the chart since its generation figures dwarf those of the other countries. From the figure, it is clear that the majority of generation in the region comes from hydropower and fossil fuels. Except for Thailand's high reliance on natural gas, an increasing percentage of which is imported, most of the fossil fuel electricity generation elsewhere comes from coal-fired plants. As shown in Fig. 10.2, Yunnan alone far outpaces its nation-state neighbours in electricity production, even though its electricity consumption is on par with that of Thailand and Vietnam.

Whereas all countries in the region have seen consistent increases in electricity demand and supply in recent years, Cambodia has seen the

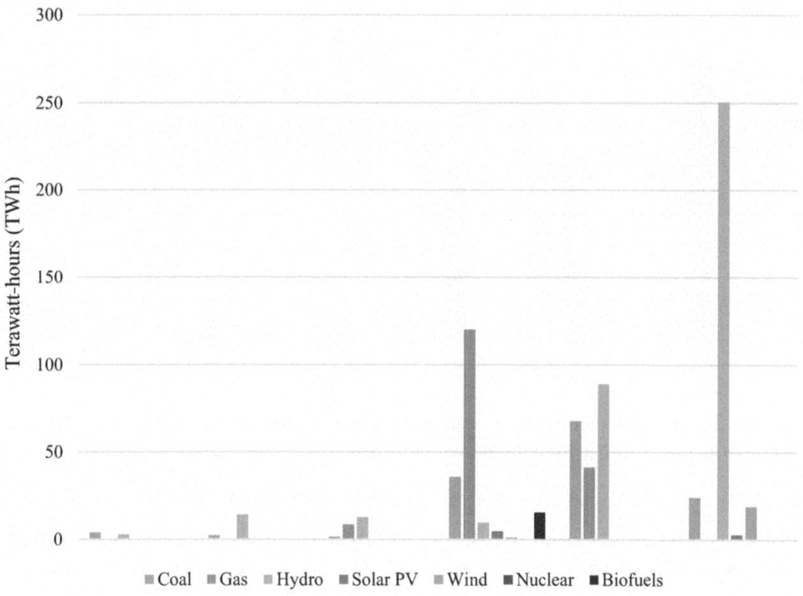

Fig. 10.1 Electricity generation by source in Yunnan and the lower Mekong region (*Source* All data are from 2017 except for Laos, which are from 2015 [Ministry of Energy and Mines, Lao PDR 2018; Yunnan Provincial Power Industry Association 2018; International Energy Agency 2020]. Data for categories "Petroleum" and "Other" not included because of their small relative values [less than 1 TWh])

most significant changes; the country's share of diesel-fired generation, especially in rural areas, has fallen, while coal-fired electricity, hydropower, and solar photovoltaic (PV) have risen. According to a report by the Electricity Authority of Cambodia, the country's total generation capacity at the end of 2018 stood at 2650 MW, a tiny figure compared to that of its neighbours, but more than a 12-fold increase domestically over the prior 15 years (Electricity Authority of Cambodia 2018). The same report noted that only 96 villages of a total of 14,168 lacked access to an electricity distribution network and did not yet have a specific timeline for obtaining such a connection.

It is not surprising that hydropower should play such an important role in electricity provision across the Lower Mekong countries, given the

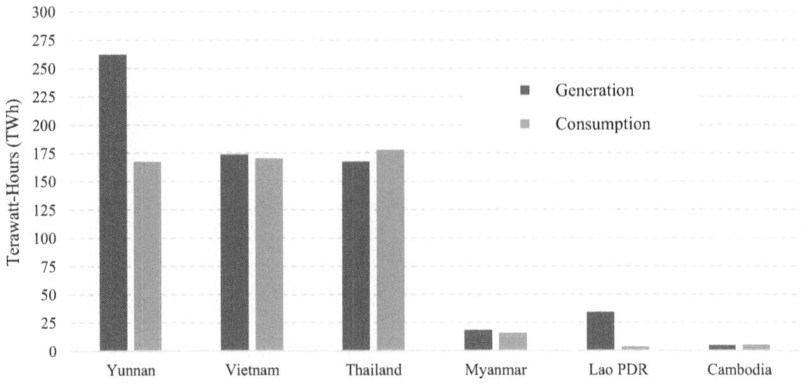

Fig. 10.2 Annual electricity consumption and generation figures for the five GMS countries and Yunnan Province (China)

density of river networks in the region and the volume of runoff generated, especially during the monsoon season. Laos, along with Yunnan Province, has styled itself "the battery of Southeast Asia" due to its sizeable hydropower potential. Both Laos and Yunnan benefit not only from heavy runoff volumes in the Lancang-Mekong and its tributaries, but also from steeper gradients than elsewhere in the lower parts of the basin. As a result, many Lao authorities are keen to develop hydroelectric dams on the Mekong's main stem as well as its tributaries in order to jump-start the Lao economy. Their reasoning is sound; according to one report, Laos exported over one billion US dollars (USD) worth of electricity in the first three quarters of 2017 (Lovgren 2018), a year when the country's nominal GDP stood at just over 17 billion USD (International Monetary Fund 2019). The Chinese dam industry, with its outward-oriented outlook and the backing of Xi Jinping's Belt and Road Initiative, is all too willing to help. In neighbouring Thailand, sentiment against further hydropower development on heavily dammed Thai rivers came to a head in the 1990s, pushing Thai authorities to look to Yunnan and Laos for hydroelectricity imports. Thailand initially signed on to a joint venture to develop Nuozhadu and Jinghong, two main-stem dams on the Lancang in Yunnan (Magee 2006a). While that initial agreement changed somewhat between the late 1990s and the time the two dams were completed, Thailand still purchases a share of its power from Yunnan and has repeatedly expressed interest in doing the same from Laos.

Despite decades-old hydropower surveys conducted in the 1950s and 1960s by American experts in the Mekong basin, downstream countries for a long time resisted the siren call to dam the Mekong's main channel. Things changed, however, after dam development took off on the upstream Chinese half of the river in the mid-1990s, first with the completion of the Manwan Dam in 1994, followed soon thereafter by the Dachaoshan, Xiaowan, Jinghong, Nuozhadu, Ganlanba, and Gongguoqiao Dams of the so-called lower Lancang cascade (Magee 2012). Still others on the upper Lancang in Yunnan and Tibet are either planned or currently underway. A database maintained by Open Development Mekong (2016) lists more than 1000 dams large and small as proposed, planned, under construction, or operational for the Mekong and its tributaries throughout the basin, a staggering number. Hecht et al. (2019) cite a more modest figure of 187 dams, but that figure only includes dams with installed capacity greater than 15 MW.

Southwestern China's tremendous hydropower potential arises from its steep mountains and big rivers. Estimates of the theoretical potential installed capacity of the five southwestern provinces and regions (Yunnan, Guizhou, Guangxi, Sichuan, and Tibet) range as high as 490 GW, equivalent to roughly 71% of the entire country's potential (Shuai 2017); not all of that potential, however, is technically or economically feasible to develop. The push to develop major hydropower cascades in western China gained real traction under the Great Western Development Campaign (*xibu dakaifa*), instituted in 1999 (Magee 2006b). Since then, the policy has only gained further momentum, with Guizhou and Sichuan exporting roughly 20% of their electricity to China's eastern and southeastern regions, and Yunnan doing so with over 30% of its own (Shuai 2017). According to an industry source, Yunnan's installed generation capacity has increased 50-fold since 1999 (Ma and Zhao 2019); another puts that increase at 71-fold since the beginning of the Reform and Opening period in 1978.

More recently, Xi Jinping unveiled his signature global infrastructure and investment programme, the Belt and Road Initiative. China's universities have trained an army of hydraulic engineers, and companies like Sinohydro, China Three Gorges Project Corporation, and Huaneng have become roving evangelists for large dams in Southeast Asia and around the world (International Rivers 2012). In mountainous Yunnan, where major Asian rivers—the Yangtze, Yellow, Mekong, Red, Irrawaddy, and Brahmaputra—come tumbling off the Tibetan Plateau through steep gorges, dam engineers see infinite potential, or at least several dozen Gigawatts of it.

All energy infrastructure carries costs, and in the Mekong basin, one of the world's richest fish- and rice-producing regions, further development of hydropower will have negative and synergistic impacts on food production and livelihoods (Eyler 2019; Hecht et al. 2019). Pittock et al. (2016: 8), for instance, note that "rapid expansion of hydro-electricity production in river basins like the Amazon, Mekong and Yangtze is having perverse impacts on biodiversity and food supply". Researchers and some policymakers refer to such synergies as occurring at the food-water-energy nexus (Foran 2015; Hennig 2016; Pittock et al. 2016). Yet government bureaucracies and the policy-making apparatus are generally not well equipped to factor in cross-sectoral impacts. Indeed, in cases where different industries and their related bureaucracies have conflicting or competing interests, incentives for considering cross-sector impacts on food production or freshwater resources, for example, might be limited or ignored. Though this is true of domestic decision-making the world over, failure to consider nexus impacts resulting from a lack of integrated planning is all but inevitable in the case of crucial transboundary watersheds such as the Mekong, where transboundary institutions are weak and geopolitical mistrust high. Perhaps nowhere else, however, is the need for collaborative, coordinated, cross-sectoral, and progressive thinking more urgent.

Balancing Variable Renewables with Hydropower

The appeal of wind and solar power is undeniable, as they represent energy sources that are infinite and free for the taking. Their ongoing operation and maintenance costs (O&M) are low and insulated from the volatility of global hydrocarbon markets. In the United States, the Levelized Cost of Electricity (LCOE) for solar PV and onshore wind is lower or essentially the same as all other options, with hydropower

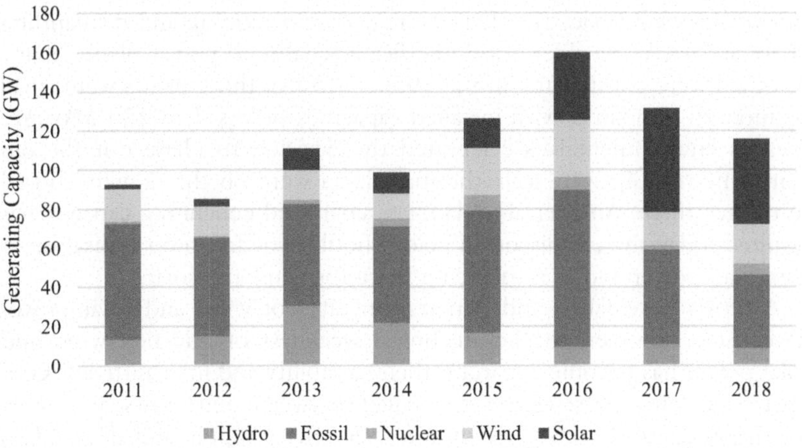

Fig. 10.3 Generating capacity additions in China from 2011 to 2018 (*Source* Hydro, fossil, nuclear, and wind data are from China Electricity Council annual summary data published in December [China Electricity Council 2011–2018]. Solar data for 2010–2017 and wind data are from the International Renewable Energy Agency [2020]. Solar data for 2018 are from National Energy Administration [2019])

close behind (US Energy Information Administration 2020).[3] Their short construction lead time and low risk makes them attractive to private investors, governments, and insurers alike. And given their rapid penetration into markets around the world, they represent an important sector for employment growth and technological advancement in the coming decades.

Figure 10.3 shows the trend in China's annual power capacity additions from 2005 to 2015. Even though the data reveal a significant increase in non-fossil generation over that period, there is also a worrisome uptick in new fossil-fuel generating capacity from 2013 to 2015. Perhaps even more striking is that those fossil fuel capacity additions occurred on the heels of the highly publicised Promote Big, Squash Small

[3] LCOE is a commonly used energy metric that takes into account construction costs, O&M costs, and other factors to facilitate cost comparisons among various electricity generating sources. While the LCOE of onshore wind, geothermal, solar PV, and hydropower are lowest, that of offshore wind is the highest of all sources considered.

(*shangdayaxiao*) policy that led to the decommissioning and dismantling of some 70 GW of older, more inefficient coal-fired power plants, many of which were constructed in the 1950s. Most of those plants were small by modern standards, with installed capacities of less than 250 MW. Yet even as environmentalists celebrated the decision to phase out the old plants, newer Gigawatt-scale thermal plants were on the drawing board to replace them. An over-abundance of coal-fired generating capacity has resulted today that, as discussed below, hobbles hydropower utilisation as base-load power and as a balancing force for wind and solar.

As costs have fallen and penetration rates of wind and solar power have risen, however, the key technical weakness of grid-tied wind and solar power has become apparent: their variability and intermittency. Grid operators, who choreograph the ballet between generators and users on the grid, must constantly monitor electricity supply and demand to ensure they are closely matched in real time. Too much or too little supply can lead to grid instability, brownouts or blackouts, economic losses, and even risk of injury. For much of its first decade of rapid wind development, China suffered particularly low levels of utilisation of its rapidly-growing wind generation assets, much of which was due to limited grid connectivity, poor planning and turbine siting, and unreliable domestically-produced turbines (Lewis 2016).

Since their output depends on resources out of the control of humans, solar and wind power are non-dispatchable; that is, their output cannot be ramped up to meet spikes in demand or to compensate for the loss of another generator due to a storm or technical failure. As such, wind and solar must be balanced (or firmed) by dispatchable resources such as gas-fired turbine generators, hydroelectric dams, or chemical batteries. Hydroelectric generators performing this role may be conventional dams on rivers or pumped storage facilities, in which water is pumped uphill from a lower reservoir at night during times of surplus electricity and allowed to flow back downhill during high consumption times during the day. Such arrangements allow grid operators to compensate for dips in output from wind and solar facilities on short notice—a lull in the wind or a passing bank of clouds—by calling on dispatchable generators to ramp up their output proportionately.

In countries or regions where so-called smart grid assets exist, communication among various components of the electrical grid can occur largely automatically and with minimal human intervention. Sensors at a solar farm, detecting a sharp drop in output caused by a passing cloud, can essentially signal to a turbine or hydroelectric facility to increase its output correspondingly until the cloud passes. To a limited extent, variable renewables can provide their own balancing if the geographic area over which the balancing occurs is large enough—that is, if the distance between solar farms or wind farms is such that a temporary drop in wind or sunlight will not affect both facilities (Apt and Jaramillo 2014: Ch. 12). In Denmark, for example, smart grid infrastructure and a cellular grid designed specifically to enable greater penetration of variable renewables also aggregates smaller, distributed generation facilities, including wind and solar, into "virtual power plants" that can mitigate the intermittency of individual wind or solar farms (Lovins and Rocky Mountain Institute 2011: 209).

Hydropower's Role in Enabling Wind and Solar Power in China

As noted previously, hydropower capacity in southwestern China is currently overbuilt and under-utilised. Shuai (2017) notes that the *qishui* (abandonment, or spillage) problem in southwestern China's hydropower sector has been growing since the start of the twenty-first century, the period over which that region has seen the most rapid buildout of large hydropower assets. Citing industry figures, Shuai indicates that the average annual spillage from southwestern hydropower facilities during the Thirteenth Five-Year Plan (2016–2020) will likely not fall below 20 TWh, a figure roughly equivalent to one-quarter of the annual output of the Three Gorges Dam, or the entire annual electricity consumption of Azerbaijan (Energy Information Administration, n.d.).

According to Shuai (2017), the principal reasons for ongoing wasted hydropower in the Southwest are over-investment in hydropower since the Twelfth Five-Year Plan, shortcomings in grid development, a slowdown in electricity consumption that coincides with an uptick in hydroelectric capacity, and deeper structural issues within the electric power system. As an example of dam construction outpacing grid development, Shuai notes that Xiluodu and Nuozhadu, two massive dams on the Jinsha and Lancang, respectively, can be forced to spill (waste) the equivalent of 6 TWh of hydroelectricity during the rainy season, since the power lines simply do not have the capacity to handle the large quantities of electricity those two dams can produce during periods of especially high runoff. Moreover, due to technical limitations in the interlinkages between China

Southern Grid and China State Grid, excess electricity production from Yunnan, covered by the Southern Grid, might occur during summertime even as Chongqing suffers blistering heat due to electricity shortages on the National Grid to which it is connected.

More broadly, there are other real challenges to retooling China's massive hydropower installed capacity—now nearly 350 GW, or twice as much as the United States' second-largest capacity—to operate as a backstop for increased wind and solar capacity. The first is that China has, and will continue to have for decades, a significant surplus of coal-fired power, some of which can be used as so-called spinning reserves to firm intermittent renewables.[4] New coal plants are still being built in China; most, if not all, are among the most efficient fossil fuel plants in the world, and can expect a minimum 40-year lifespan.[5] The implication for the hydropower and variable renewables sub-sectors is that in some regions, at least, there may be little demand for hydropower as a balancing or firming technology for solar and wind as long as coal- and gas-fired power plants are operating under capacity and can be tapped to fill dips in wind and solar output. The principal obstacles to using thermal plants in this way are that they have slower ramp rates than hydroelectric dams, and that ramping their output up and down can cause increased thermal stress on plants (thus increasing O&M costs) as well as increased carbon emissions from less-efficient operation.

The second challenge is that dispatch decisions—that is, decisions about which generators are called upon at what points and for how much output—are at once technical, political, and economic decisions (Cheng et al. 2019). To the extent that high-carbon emitters are penalised (either through taxes or market mechanisms), hydropower's role in balancing will likely increase. That has traditionally not been the case in China,

[4] Spinning reserves are power plants, usually gas- or coal-fired, that are kept idling during periods of high intermittent renewable output in case that output falls. The spinning reserve systems can then be ramped up to compensate for a drop-in renewable output more quickly and safely than if they had completely powered down.

[5] One might just as well say the 'least inefficient' here, since even the best thermal power plants in the world still dump between half and two-thirds of their energy to the environment as "waste heat" due simply to the laws of thermodynamics and the limitations of materials engineering. The only exception to this rule are combined heat and power (CHP) plants that put "waste heat" to productive use heating nearby buildings or as process heat in factories.

however, where coal-fired power plants—even the oldest and most inefficient ones—have long been allocated generation quotas to maintain high employment levels in the plants and in the coal supply chains feeding those plants. Reforms introduced in 2015, however, now prioritise "new" renewable sources including wind and solar (Liu and Kong 2016). These same reforms also call for gradually introducing pricing reforms and could provide a pathway for large purchasers of electricity to enter into direct power purchase agreements (PPAs) with generators. Wind, solar, and hydropower, with their lower emissions and continuously declining production costs, will almost certainly benefit from greater utilisation due to these reforms, assuming they are fully implemented.

Changing the "software" of China's power sector—that is, the operational rules of the electricity market—is not a new idea. Ever since reforms at the turn of the twenty-first century separated grid operators from generators (*chanwang fenkai*), control over the distribution of electricity once it leaves the generator has become the purview, theoretically at least, of the grid operator—China Southern Power Grid in the case of Yunnan and Guizhou, National Grid in the case of Sichuan. Control of electricity prices, meanwhile, rests with the central government. Establishing a wholesale market for electricity, where generators and grid operators can set prices for electricity to be delivered in the immediate future ('spot' market), hours ahead, or days ahead, will enable hydropower operators, for example, to use pricing to signal periods when they expect to have a shortage or surplus of water in their reservoirs. Experts in the wind sector also recognise the value of using pricing rather than quotas to drive dispatch decisions (Winglee 2016), a system that could readily take into account factors such as carbon taxes or incentives to purchase electricity from carbon-free generation sources and which would make dispatch decisions based on criteria other than maintaining employment at coal-fired facilities.

A third significant set of challenges to improving dispatch for the hydropower sector lies in the sheer complexity of the hydropower and grid landscape in southwestern China. As explained by Cheng et al. (2019) in a recent article on the challenges of improving dispatch in the era of Gigawatt-scale dams, these include simple problems of scale: southwestern China's rivers are already home to some 70% of the installed hydropower capacity in the entire China Southern Grid. Most of the mainstream dams have very large reservoirs, high hydraulic heads, and are part of cascades numbering 10–20 dams along one river. In such large cascade arrangements, changes in the operating parameters of one dam—to meet peaking power needs, for instance—will affect the hydraulic head and reservoir volumes of not only that dam but also other dams downstream. As a consequence, the flexibility of those downstream dams may then be constrained in unexpected ways.

Building or upgrading transmission lines to connect widely dispersed generation facilities is expensive, however, and the costs of doing so would have to be weighed against other options. This is especially true in places such as China where a lack of carbon pricing or strong vested interests in fossil fuels make balancing with fossil-fuel generators more economically viable. From early 2018 to mid-2019, for instance, China saw a net increase of 49.2 GW of new coal-fired generation capacity (Ambrose 2019), a figure roughly equivalent to the total installed generating capacity of Turkey or Saudi Arabia. Much of that fossil fuel generating capacity remains under-utilised, however, meaning that it could relatively easily provide so-called spinning reserves as a hedge against potential drops in renewable power output.

Cheng et al. (2019) also note other characteristics in the Southwest's hydropower sector which, if not unique to the region, are at least uncommon and lack reference models in China or elsewhere. Significant seasonal variations in runoff, for instance, can overwhelm small- to mid-size hydropower stations on tributaries and lead to spillage as grid assets become clogged with too much hydroelectricity.

Going Outward: Using Yunnan Hydropower to Enable a Solar Transition in Cambodia

Charting a future electricity supply paradigm for the Mekong region that is less reliant on large-scale hydropower development in the lower basin could increase food supply security and take advantage of abundant non-hydropower renewable resources in the region (Pittock et al. 2016). Cambodia, in particular, is a solar bright spot in the Mekong region. The country's levels of solar irradiance are among the highest in the world at 5 kWh per m^2 per day, comparable to those of the Sahara Desert (World Bank Group et al. 2020).[6] Based on that fact alone, an area of

[6] See "Global Solar Atlas" (World Bank Group et al. 2020). Data for the atlas come from the SolarGIS at https://solargis.info/.

solar panels roughly one-fifteenth the size of Cambodia's iconic Tonle Sap Lake during the dry season (183 km^2), could theoretically supply enough electrical energy to meet the country's current electricity needs. As noted above, however, Cambodia's electrical system at present is largely powered by fossil fuels, and even though the declining share of diesel generators in Cambodia's electricity mix is a positive sign, the coal-fired power plants replacing such generators are hardly emissions-free. In fact, while nearby China's new Gigawatt-scale coal-fired power plants are among the most efficient in the world, Cambodia's nascent fleet of coal-fired power plants are smaller facilities with yesterday's technologies, resulting in lower efficiency and higher emissions.

Unreliable electricity availability, combined with high tariffs for that electricity, has literally left many rural Cambodians in the dark for much of the last half-century. As depicted in Fig. 10.4, the percentage of the population of Cambodia and Laos with access to electricity increased from 2010 to 2016, with levels still quite low among rural Cambodians. As noted earlier, however, Cambodia's Electricity Authority claimed that, by the end of 2018, only 96 villages lacked grid connections, suggesting a good deal of progress was made in the intervening two years. Scholars of Cambodia's electricity system have long been aware of this problem

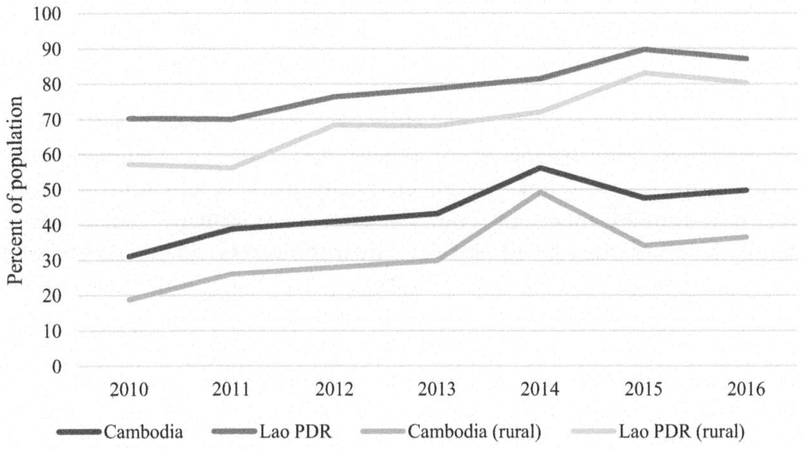

Fig. 10.4 Access to electricity in Laos and Cambodia (*Data source* World Bank [2020])

and have argued that renewable energy should play a significant role in meeting Cambodia's electricity needs, some 5 billion kWh annually. Some of those might be conventional renewable energy sources, such as solar PV, which have found strong policy support for nearly a decade now (Sarraf et al. 2013). Others might be more specifically related to the particularities of Cambodia's economy, such as using rice husks to power thermal power plants or, through anaerobic digestion of the husks, producing methane that could then power gas-fired thermal power plants (Pode et al. 2015). By virtue of its wartime history, which left much of the country's rural infrastructure decimated, including its power grid, the country is ripe for distributed renewables as a solution for rural electrification.

Yet given the intermittency of solar power described above, there is some benefit to imagining a grid design that, while incorporating micro-grids that take advantage of localised resources such as rooftop solar and micro-hydropower, could at the same time benefit from interconnectivity with a regional grid. Such a grid, on the minds of GMS planners since the first days of the GMS itself, would allow utility-scale solar power assets (and wind, to the extent those are developed) to benefit from the firming capabilities of Yunnan's highly developed hydropower sector. In such a scenario, smart grid technologies would allow near-instantaneous communications between solar farms in Cambodia and hydropower stations in Yunnan, so that in the event of a drop in solar output, Yunnan's hydropower could quickly and automatically make up the difference.

There already exists a model for such transboundary grid interconnectedness and balancing of renewables: Norway, Denmark, and Germany have tightly linked their electricity markets and grids so as to take maximum advantage of renewable sources—Germany's solar power, Denmark's wind power, and Norway's hydropower. Germany has invested heavily in solar (and wind) installations since the Fukushima nuclear accident in 2011, when longstanding scepticism among many of Germany's citizens and leaders coalesced around a complete phase-out of nuclear power by 2022. Since 2011, Germany, roughly the size of Yunnan, has partly made up for its declining use of nuclear power by aggressively investing in solar and wind power, with capacity doubling from 25 to 49 GW in solar power and 28 GW to 60 GW in wind power

(Fraunhofer ISE 2020).[7] Denmark, meanwhile, drastically reduced its coal-fired electricity production from 40 TWh in 1996 to about 7 TWh in 2018, while growing its wind electricity generation from 1227 TWh to 13,899 TWh (International Energy Agency 2020). Finally, Norway obtains roughly 95% of its generated electricity from hydropower stations, providing a reliability guarantee to counter the inevitable intermittency of Danish wind and German sun. In the Mekong region, Yunnan could in a similar fashion provide the reliability guarantee to balance intermittent renewables downstream. Cambodia, with far higher insolation levels than Germany, completed its first utility-scale solar farm in Svay Rieng Province in late 2017 (Asian Development Bank, n.d.). Funding for a new 100 MW project was recently approved by the Asian Development Bank, whose experts noted that "Cambodia can add about 200 MW of solar energy to the grid by 2021, using available technology and without disrupting the grid" (Asian Development Bank 2019: online). The Bank's concerns about grid disruption stem from the intermittency problem of large-scale solar farms, which, I have argued, could be mitigated through better integration of Yunnan's surplus hydropower. To be sure, resolving hardware issues such as limited grid interconnectivity and long-distance transmission lines, as well as software issues such as the role of Laos as a transmission intermediary, will not be a trivial affair.

Conclusion

Since the beginning of the reform period, China's electricity demand has grown in lock-step with the country's GDP. Against this backdrop, the primary goal of the electric power regulatory apparatus was, as Liu and Kong (2016: 2) note, "to ensure sufficient electricity supply at acceptable prices" for the engines of China's export-oriented manufacturing economy. It is little wonder, then, that so much capital for power infrastructure development during the first decade of the twenty-first century focused on directly connecting energy-hungry provinces like Guangdong to energy-producing provinces like Yunnan. Now, as China's economy slows and the pressure to power a rapidly-growing industrial sector eases, policymakers may be emboldened to take greater steps towards liberalisation of the electricity market and letting supply and demand

[7] See "Net installed electricity generation capacity in Germany" at Energy Charts (https://www.energy-charts.de/power_inst.htm), accessed 16 June 2020.

dynamics have a greater effect on pricing. China's electricity market reforms launched in 2015 included pricing reform projects in Shenzhen, Inner Mongolia, Hubei, Anhui, Yunnan, Guizhou, and Ningxia (Liu and Kong 2016). Enabling greater flexibility in electricity pricing, and in the contractual arrangements among generators, transmission operators, and end users, would allow generators to command higher prices for factors such as availability, low carbon emissions, or frequency regulation, while empowering buyers to take advantage of those characteristics as needed.

At the same time, China's leaders continue to push for greater levels of electrification in the consumer sector, especially in rural areas. Some of these policies may appear to target efficiency, but they are equally designed to increase rural consumption of goods such as electrical appliances, therefore serving the dual purposes of increasing domestic consumption of manufactured goods as exports decline and increasing electricity utilisation.

China's favourable geography, massive hydropower assets, and rapidly growing network of communications-enabled smart grid assets all point to a power sector future—not only in China, but throughout the region—in which the technical challenges of widespread deployment of wind and solar farms are increasingly less problematic. Yet there remains much carbon on China's horizon due to the continued buildout of new fossil fuel power plants and the lack of political will to begin dismantling the coal industry and risk the social upheaval that will likely follow as millions of workers are dislocated. It is easy to envision a scenario where hydropower producers team up with wind or solar producers, leveraging hydropower's ability to reduce intermittency alongside policies that incentivise grid operators and major electricity users to purchase those "new energy" sources both within China and throughout the Mekong region. The technology exists, as does the readiness among development organisations such as the Asian Development Bank, which has pushed region-wide grid development for nearly two decades. What remains are the innovations in policies and markets both domestically and across boundaries that can facilitate the changes in networks of power—literal and figurative—necessary for a more sustainable energy future in the Mekong region.

References

Ambrose, J. (2019, November 20). China's appetite for coal power returns despite climate pledge. *The Guardian*. https://www.theguardian.com/world/2019/nov/20/china-appetite-for-coal-power-stations-returns-despite-climate-pledge-capacity. Accessed 16 June 2020.

Apt, J., & Jaramillo, P. (2014). *Variable renewable energy and the electricity grid*. New York: RFF Press.

Asian Development Bank. (2019, May 24). *ADB to help build 100-MW solar park in Cambodia*. https://www.adb.org/news/adb-help-build-100-mw-solar-park-cambodia. Accessed 16 June 2020.

Asian Development Bank. (n.d.). *Cambodia: Cambodia solar power project*. https://www.adb.org/projects/50248-001/main. Accessed 16 June 2020.

Brown, P., Tullos, D., Tilt, B., Magee, D., & Wolf, A. (2009). Modeling the costs and benefits of dam construction from a multidisciplinary perspective. *Journal of Environmental Management, 90*(Suppl. 3), S303–S311.

Cheng, C., Wu, X., Shen, J., Li, G., Liao, S., & Liu, B. (2019). A state-of-the-art review of China's hydropower operations and the recent advances in the era of gigawatts. *Journal of Water Conservancy, 50*(1), 112–123.

China Electricity Council. (2011–2018). *National electric power industry statistical data* (January–November). https://www.cec.org.cn/guihuayutongji/tongjxinxi/yuedushuju/. Accessed 16 June 2020.

Electricity Authority of Cambodia. (2018). *Salient features of power development in Kingdom of Cambodia until December 2018*. https://eac.gov.kh/uploads/salient_feature/english/salient_feature_2018_en.pdf. Accessed 16 June 2020.

Energy Information Administration. (n.d.). *International energy statistics: Total electricity net consumption 2017*. https://www.eia.gov/beta/international/rankings/#?iso=KHM&cy=2017&aid=2&pid=2&tl_id=2-A. Accessed 4 December 2019.

Eyler, B. (2019). *Last days of the mighty Mekong*. London: Zed Books.

Foran, T. (2015). Node and regime: Interdisciplinary analysis of water-energy-food nexus in the Mekong Region. *Water Alternatives, 8*(1), 655–674.

Fraunhofer ISE. (2020). *Energy charts: Net installed electricity generation capacity in Germany*. https://www.energy-charts.de/power_inst.htm. Accessed 16 June 2020.

Goldsmith, E., & Hildyard, N. (1986). *The social and environmental effects of large dams*. San Francisco: Sierra Club Books.

Hecht, J. S., Lacombe, G., Arias, M. E., Dang, T. D., & Piman, T. (2019). Hydropower dams of the Mekong River basin: A review of their hydrological impacts. *Journal of Hydrology, 568*, 285–300.

Hennig, T. (2016). Damming the transnational Ayeyarwady basin. Hydropower and the water-energy nexus. *Renewable and Sustainable Energy Reviews, 65*, 1232–1246.

Hennig, T., Wang, W., Feng, Y., Ou, X., & He, D. (2013). Review of Yunnan's hydropower development. Comparing small and large hydropower projects regarding their environmental implications and socio-economic consequences. *Renewable and Sustainable Energy Reviews, 27,* 585–595.

Ho, M., Lall, U., Allaire, M., Devineni, N., Kwon, H. H., Pal, I., et al. (2017). The future role of dams in the United States of America. *Water Resources Research, 53*(2), 982–998.

International Energy Agency. (2020). *Data and statistics*. https://www.iea.org/data-and-statistics. Accessed 16 June 2020.

International Monetary Fund. (2019). *World economic outlook database.* https://www.imf.org/external/pubs/ft/weo/2019/02/weodata/index.aspx. Accessed 16 June 2020.

International Renewable Energy Agency. (2020). *Data download*. https://resourceirena.irena.org/gateway/dashboard/?topic=4&subTopic=16. Accessed 16 June 2020.

International Rivers. (2012). *The new Great Walls: A guide to China's overseas dam industry*. https://www.internationalrivers.org/resources/the-new-great-walls-a-guide-to-china%E2%80%99s-overseas-dam-industry-3962. Accessed 16 June 2020.

International Rivers. (n.d.). *Irrawaddy Myitsone Dam*. https://www.internationalrivers.org/campaigns/irrawaddy-myitsone-dam-0. Accessed 16 June 2020.

Kibler, K. M., & Tullos, D. D. (2013). Cumulative biophysical impact of small and large hydropower development, Nu River China. *Water Resources Research, 49*(6), 3104–3118.

Lewis, J. I. (2016). Wind energy in China: Getting more from wind farms. *Nature Energy, 1*(6), 16076.

Liu, X., & Kong, L. (2016). *A new chapter in China's electricity market reform* (ESI Policy Brief 13). https://esi.nus.edu.sg/docs/default-source/esi-policy-briefs/a-new-chapter-in-china-s-electricity-market-reform.pdf. Accessed 16 June 2020.

Lovgren, S. (2018, August 23). Southeast Asia may be building too many dams too fast. *National Geographic*. https://www.nationalgeographic.com/environment/2018/08/news-southeast-asia-building-dams-floods-climate-change/. Accessed 16 June 2020.

Lovins, A. B., & Institute, R. M. (2011). *Reinventing fire: Bold business solutions for the new energy era*. White River Junction, VT: Chelsea Green Pub.

Ma, S., & Zhao, Z. (2019). *Yunnan grid installed capacity surpasses 90 Gigawatts*. https://www.chinapower.com.cn/dwsj/20190820/1284135.html. Accessed 16 June 2020.

Magee, D. (2006a). *New energy geographies: Powershed politics and hydropower decision making in Yunnan, China* (PhD dissertation). University of Washington.

Magee, D. (2006b). Powershed Politics: Yunnan hydropower under Great Western Development. *The China Quarterly, 185*, 23–41.

Magee, D. (2012). The dragon upstream: China's role in Lancang-Mekong development. In J. Öjendal, S. Hansson, & S. Hellberg (Eds.), *Politics and development in a transboundary watershed: The case of the Lower Mekong basin* (pp. 171–194). New York: Springer.

Magee, D. (2014). Dams in East Asia: Controlling water but creating problems. In P. Harris & G. Lang (Eds.), *Routledge handbook of environment and society in Asia* (pp. 216–236). London: Routledge.

Magee, D., & Hennig, T. (2017, April 28). Hydropower boom in southwestern China outpaces regional electricity demand. *The Third Pole*. https://www.thethirdpole.net/2017/04/28/hydropower-boom-in-china-and-along-asias-rivers-outpaces-regional-electricity-demand/. Accessed 16 June 2020.

McCully, P. (1996). *Silenced rivers: The ecology and politics of large dams*. New York: Zed Books.

Ministry of Energy and Mines, Lao PDR. (2018). *Lao PDR Energy Statistics 2018*. Economic Research Institute for ASEAN and East Asia. https://www.eria.org/uploads/media/0_Lao_PDR_Energy_Statistics_2018_complete_book.pdf. Accessed 16 June 2020.

National Energy Administation. (2019, January 18). *National Energy Administration releases 2018 national electric power industry statistical data*. https://www.nea.gov.cn/2019-01/18/c_137754977.htm. Accessed 16 June 2020.

Open Development Mekong. (2016). *Greater Mekong Subregion hydropower dams (2016)*. https://data.opendevelopmentmekong.net/en/dataset/greater-mekong-subregion-hydropower-dams-2016. Accessed 16 June 2020.

Pittock, J., Dumaresq, D., & Bassi, A. M. (2016). Modeling the hydropower–food nexus in large river basins: A Mekong case study. *Water, 8*(10), 425.

Pode, R., Diouf, B., & Pode, G. (2015). Sustainable rural electrification using rice husk biomass energy: A case study of Cambodia. *Renewable and Sustainable Energy Reviews, 44*, 530–542.

Poff, N. L., & Hart, D. D. (2002). How dams vary and why it matters for the emerging science of dam removal: An ecological classification of dams is needed to characterize how the tremendous variation in the size, operational mode, age, and number of dams in a river basin influences the potential for restoring regulated rivers via dam removal. *BioScience, 52*(8), 659–668.

Sarraf, M., Rismanchi, B., Saidur, R., Ping, H. W., & Rahim, N. A. (2013). Renewable energy policies for sustainable development in Cambodia. *Renewable and Sustainable Energy Reviews, 22*, 223–229.

Scudder, T. (2005). *The future of large dams: Dealing with social, environmental and political costs*. Sterling, VA: Earthscan.

Shuai, W. (2017). The problems and countermeasures about abandoned water in China Southwest hydropower station. *Electric Power, 50*(10), 171–175. https://doi.org/10.11930/j.issn.1004-9649.20160012
US Energy Information Administration. (2020). *Levelized cost and levelized avoided cost of new generation resources in the annual energy outlook 2019.* https://www.eia.gov/outlooks/aeo/pdf/electricity_generation.pdf. Accessed 16 June 2020.
Winglee, M. (2016, July 22). How China can stop wasting wind energy. *China Dialogue.* https://www.chinadialogue.net/article/show/single/en/9119-How-China-can-stop-wasting-wind-energy. Accessed 16 June 2020.
World Bank. (2020). *Data: Access to electricity.* https://data.worldbank.org/indicator/EG.ELC.ACCS.ZS. Accessed 16 June 2020.
World Bank Group, SolarGIS, Energy Sector Management Assistance Program. (2020). *Global Solar Atlas 2.2.* https://globalsolaratlas.info. Accessed 16 June 2020.
World Commission on Dams. (2000). *Dams and development: A new framework for decision-making.* London: Earthscan.
Yunnan Provincial Power Industry Association. (2018, January 20). *Important indicator operational for Yunnan power sector January–December 2017.* https://www.yppta.com.cn/pc/front/cms/articleDetail?articleId=8844c1ba8a6011e8a1a7f6ecbce650db&categoryId=13f748a07dd111e88dc400163e003558. Accessed 16 June 2020.

CHAPTER 11

China's Hydro-Hegemony in the Mekong Region: Room for Improvement

Sebastian Biba

Introduction

Freshwater resources do not respect political boundaries and therefore are frequently shared among countries, making water management an issue of international concern in many places. According to the Oregon State University's Register of International River Basins, there are as many as 310 international river basins, covering 47.1% of global land surface (McCracken and Wolf 2019). Observing a worrisome trend, however, De Stefano et al. (2010) have found that the number of conflictive events involving riparian neighbours in international river basins has, overall, been on the rise, particularly regarding infrastructure and water quantity. In the light of this situation, which is further aggravated by relatively

S. Biba (✉)
Goethe University Frankfurt, Frankfurt, Germany
e-mail: biba@soz.uni-frankfurt.de

School of Advanced International Studies (SAIS), Johns Hopinks University, Washington, DC, USA

J.-F. Rousseau and S. Habich-Sobiegalla (eds.), *The Political Economy of Hydropower in Southwest China and Beyond*, International Political Economy Series, https://doi.org/10.1007/978-3-030-59361-2_11

weak international water law, good governance of shared water resources becomes more important than ever. Concomitantly, the role of the most powerful riparian countries, the so-called hydro-hegemons, also takes centre stage, because power entails responsibility.

In the Asian context, China readily comes to mind as a hydro-hegemon (Chellaney 2013; Ho 2016; Biba 2018a). China's favourable geographical position—containing the Qinghai-Tibet Plateau, where the world's largest freshwater reserves outside the polar ice caps are located and where several major transboundary rivers originate—makes it an unrivalled, multi-directional water provider for its entire periphery (Chellaney 2011). Together with the fact that China today also boasts the world's second-biggest military and economy, it is not hard to understand why Nickum (2008: 227) has referred to China as an "upstream superpower". However, such a label does not say much about the quality of China's behaviour. How has China made use of its superior geographical position and material power? Simply put, has China been a good or bad hydro-hegemon?

To answer these questions, this chapter will apply the framework of hydro-hegemony as introduced by Zeitoun and Warner (2006) to China's performance in the Mekong River region. This framework has been designed to analyse the power-related strategies and tactics of how water resources are controlled in places where power imbalances exist at the river basin level (Zeitoun and Warner 2006). The Mekong River, known as the Lancang in China, represents the most important waterway in Southeast Asia and, in addition to its source in China, flows through Myanmar, Laos, Thailand, Cambodia, and Vietnam. The Mekong River is chosen as a case study for China's behaviour mainly because the Southeast Asian states in its watershed are arguably among the weakest of China's riparian neighbours, thus making China's position as hydro-hegemon all the more pronounced. The main argument that this chapter will advance is that China's hydro-hegemonic behaviour is actually hard to classify neatly as either completely positive or entirely negative. This, however, also means that there is still room for China to improve its track record.

The remainder of this chapter will proceed in four major steps: Zeitoun and Warner's hydro-hegemony framework will first be outlined and explained in greater detail. This conceptual framework will then be applied to China's behaviour in the Mekong region, illustrating the specific hegemonic mechanisms and control strategies China has employed. The next section will explore the reactions to China's behaviour expressed by its neighbours and others beyond the region. Finally, the conclusion will present the main findings of the preceding analyses.

The Conceptual Framework of Hydro-Hegemony[1]

As those who coined the term have explained, the concept of hydro-hegemony refers to hegemony at the river basin level and its goal is to provide a (new) answer to "the perennial and deeply political question: who gets how much water, how and why?" (Zeitoun and Warner 2006: 435). At the outset of conceptual concern therefore lies the notion of hegemony and its meaning. According to the Oxford English Dictionary, hegemony means "leadership or dominance, especially by one state or social group over others". Judging from this definition, hegemony can come in two almost diametrically opposed forms—that is, one that is underpinned by authority and can be seen as positive, and another that is buttressed by coercion and can be seen as negative. This is also true in the context of hydro-hegemony. As Zeitoun and Warner (2006: 439) have elaborated, in a positive manifestation of hydro-hegemony, the hegemon's actions are "for the good of all or most under its sphere", resulting in the provision of public goods such as order, stability, and assurance of water flow. In contrast, a negative form of hydro-hegemony is built upon domination and leads to ever-growing inequalities that favour the water 'rights' of the strong over the weak.

Either way, it is any hegemon's key objective to remain in control and thus to ensure compliance. Inspired, above all, by both Realist and Gramscian ideas, Zeitoun and Warner (2006) have identified four types of compliance mechanisms a hegemon can use. The first is coercive, meaning the threat or actual use of force or diplomatic pressure. The second is utilitarian, implying the offering of incentives, mostly economic and financial. The third is normative, referring to the signing of treaties and the building of institutions that preserve the hegemon's position. Finally, the fourth

[1] Following Zeitoun and Warner's introduction of the framework of hydro-hegemony in 2006, there has been a number of attempts at revising and refining the concept, most prominent of which probably are Cascao (2008), Cascao and Zeitoun (2010), Menga (2016), and Warner et al. (2017). The purposes of this chapter are sufficiently served by applying key ideas of the original concept, though.

is ideological, describing a process of discursive knowledge construction that not only aims for the acceptance of the hegemon's superiority, but also for the internalisation of the hegemon's values and norms by inferior actors.

That said, (hydro-)hegemony is intimately linked to another essential concept: power. This is not only because the state with the largest aggregate power resources qualifies as the hegemon; it is also because the toolbox of a hegemon—that is, which of the four compliance mechanisms a hegemon can make use of—depends on what kind(s) of power it is able to wield. Zeitoun and Warner (2006) incorporate various notions of power (cf. Dahl 1965; Strange 1987; Lukes 2005) to end up with three different dimensions that combine both so-called hard and soft forms of power (also see Cascao and Zeitoun 2010; Menga 2016). The first dimension is material power, which particularly includes economic might and military capabilities. The second is labelled bargaining power, referring to the capability of actors to control the rules of the game and set agendas. The third is understood as ideational power, or power over ideas, and represents the capacity to impose and legitimise ideas and narratives. Evidently, these different dimensions of power can be linked to the four hegemonic compliance mechanisms: material power is linked to the coercive and utilitarian mechanisms, bargaining power is related to the normative mechanism, and ideational power is connected with the ideological mechanism.

While the three dimensions of power outlined here may be critical for hegemons in general, two additional aspects are significant in the specific context of hydro-hegemons. These are the riparian position (upstream vs. downstream) on the one hand and the exploitation potential in terms of technological capabilities and infrastructure development on the other. Together with power, they make up the three pillars of hydro-hegemony (Zeitoun and Warner 2006).[2]

With the various power sources at its disposal, a hydro-hegemon is then confronted with the question of how to reach its ultimate goal of control over water resources. According to Zeitoun and Warner (2006), there are

[2] In their advancement of the original hydro-hegemony framework, Cascao and Zeitoun (2010) frame hydro-hegemony as consisting of four kinds of power (instead of three), in which riparian position is conceived of as geographical power and exploitation potential is subsumed under material power. The other two continue to be bargaining and ideational power.

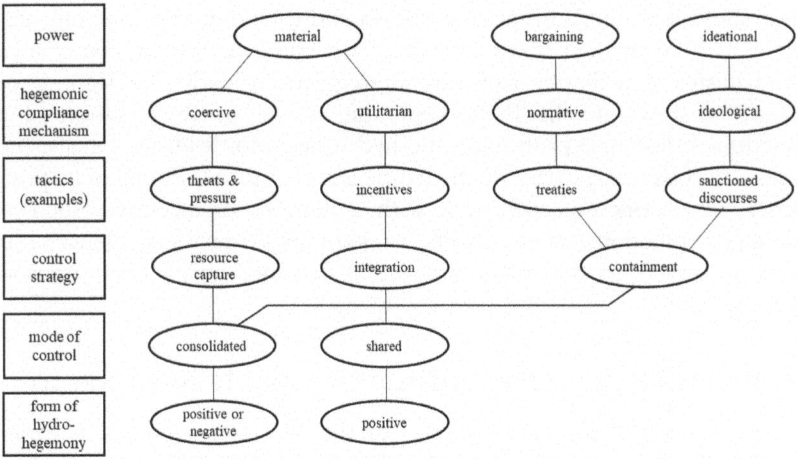

Fig. 11.1 How to arrive at the different forms of hydro-hegemony (*Source* Information from Zeitoun and Warner [2006])

two general modes of hydro-hegemonic control. One is shared control, which is cooperative with respect to riparian interaction and therefore points to a positive hegemony. The other is consolidated control, with a competitive mode of riparian interaction that can either be positive or lean towards a negative form of hegemony.[3] To achieve one of these two modes, hydro-hegemons can rely on three different control strategies: resource capture (also called active unilateralism), containment, and integration (Zeitoun and Warner 2006). These three strategies, in turn, can generally be implemented through the use of tactics related to the four hegemonic compliance mechanisms identified above (Zeitoun and Warner 2006; see also Fig. 11.1).

More precisely, the unilateral resource capture strategy, which seeks to create *faits accomplis*, may be dependent on a favourable geographical position and existing exploitation potential. In addition, this strategy may be supplemented by measures such as pressure and threats stemming from the coercive hegemonic compliance mechanism. The containment

[3] As a matter of fact, control may also be contested, but in those cases there is no hydro-hegemony or hydro-hegemony is in a state of flux (Zeitoun and Warner 2006).

strategy, meanwhile, aims at co-optation and thus requires some engagement with co-riparian states. The normative and ideological compliance mechanisms may better suit this strategy, particularly the signing of treaties that favour the hydro-hegemon as well as the creation of a sanctioned discourse promoting the hydro-hegemon's ideals. Finally, the integration strategy, where the hydro-hegemon concedes some of its privileges, may work especially well with a number of incentives such as military protection, trade benefits, or joint hydro-projects. Hence, this strategy reverts to the utilitarian compliance mechanism (Zeitoun and Warner 2006).

China's Hegemonic Choice in the Mekong Region

Before delving into China's actions in the Mekong region, it is expedient to briefly (re-)confirm China's hegemonic status there in the first place. In light of the different power-related qualities comprising a hydro-hegemon according to Zeitoun and Warner (2006), the following can be said: China is the Mekong River's most upstream riparian country and therefore holds a favourable geographical position. China also constitutes by far the most powerful country in the river basin as far as military and economic indicators are concerned. China's GDP (measured in purchasing power parity) has been estimated to have stood at 23.21 trillion US dollars (USD) in 2017, whereas the combined equivalent for the other five Mekong River riparian countries totalled only 10% of this figure (CIA 2019). Moreover, China spent an estimated 1.87% of its GDP on its military in 2018, while the other five Mekong countries ranged between 0.19% (Laos) and 2.93% (Myanmar) in this regard (CIA 2019). In absolute figures, the difference is enormous since China's GDP is much higher than the respective GDPs of the other five countries. Hence, China is clearly the country with the greatest material power resources in the Mekong basin. While bargaining and ideational power are more difficult to measure, we may assume at least a loosely proportional relationship between these types of power and material power, in that whoever possesses enormous material power is likely to also bring a large amount of bargaining and ideational power to the table (see Young 1991). Finally, the combination of China's upstream position and its superior level of economic sophistication hand it the biggest potential for exploiting the Mekong River's resources, too. In sum, then, there is no doubt that China is the undisputed hydro-hegemon in the Mekong basin. What is more, China generally speaking has the power resources at its disposal to make a conscious choice about whether it wants to be a good or bad hydro-hegemon. The following section will now analyse China's hydro-hegemonic performance in the Mekong region, discussing which of Zeitoun and Warner's (2006) three control strategies China has made use of and how.

Resource Capture Strategy

The clearest evidence that China has been engaged in a resource capture strategy in the Mekong River basin is its building of large-scale dams on the upstream stretch of the river since the mid-1980s. As of this writing, China has completed the construction of at least six large hydropower dams on the lower reaches of the Lancang River's main branch, while an even higher number of dams is currently in the building or planning stage upstream (International Rivers 2019; see Hennig and Magee, this volume).[4] These activities constitute an act of resource capture because China has constructed the dams in a strictly unilateral fashion. Notably, China has not been willing to discuss or consult on its dam building with other Mekong basin countries. In fact, China has never even informed these countries about its plans and actions in advance, but has instead presented them with *faits accomplis* (Biba 2018b).

Given that such large-scale infrastructure developments may well cause adverse effects downstream, China's unilateral activities upstream must be considered all the more problematic. As International Rivers (2019: online) has pointed out, scientific reports have found that "changes in hydrology, fisheries and sedimentation brought by the Lancang dams" are well underway and likely to "have extensive and very significant impacts on millions of people who rely directly on the river for their food and livelihoods". But that is not all. Ultimately, China's upstream dams put Beijing in a position to control the quantity of water flowing downstream. The dams thus constitute a potentially powerful instrument to exert leverage on the downstream countries and possibly pose a diplomatic threat (Biba 2012).

Recently, we were given some early insights into China's water control capacities and their implications. In summer 2019, Chinese authorities

[4] Note that due to China's tight information policy, the precise number of China's planned and completed Lancang dams varies by source.

opened the gates of the Jinghong Dam for maintenance purposes, thereby causing floods downstream in Laos and Thailand that reportedly damaged crops and fisheries. After the repair work was done, Chinese authorities refilled the emptied dam, resulting in highly reduced water levels downstream during an already very dry period (Hutt 2019).[5] Analogously, in spring 2016, China released water from its Lancang dams to mitigate a severe drought in the lower basin, following a request from Vietnam. Due to a lack of coordination, however, these water releases turned out to be so unpredictable that they reportedly had negative impacts on some riverside communities in Thailand (Biba 2018b). In other words, China's recent approach to communicating the operating schemes of its completed dams is just as opaque as its earlier behaviour with regard to building these dams in the first place and precipitates similarly negative impacts.

Last but not least, China has been reluctant to sign a water-sharing agreement with its downstream neighbours on the Mekong River. This is related to China's "absolutist stance on its sovereign rights to develop the water resources that flow through its territory" (Ho 2016: 36). Consequently, China has steadfastly rejected becoming a member of the Mekong River Commission (MRC), the Mekong's leading water resources governance mechanism, despite repeated overtures from its four member states (Cambodia, Laos, Thailand, and Vietnam). Among the many reasons for its behaviour is above all that China, in its upstream and therefore relatively invulnerable position, has seen no incentive to enter a multilateral organisation whose predefined rules would only attempt to restrict China's autonomy and freedom of action (see Biba 2018a). Instead, China in 1996 chose to only accept an observer status in the MRC that comes without any obligations attached.

Integration Strategy

Interestingly, China's hydro-politics has not only followed a resource capture strategy in the Mekong region. A case can well be made that

[5] It is also noteworthy that from April to November 2019, China was claimed to have "turned off the tap on the Mekong River" by impounding much more water than ever before, thereby driving the severe drought situation in the lower basin (Eyler and Weatherby 2020: online). Others, however, have questioned whether the satellite data available justify such a serious claim (Ketelsen et al. 2020).

China's behaviour also includes traits of the opposite strategy, that is, integration. According to Zeitoun and Warner (2006), it is not unusual for hydro-hegemons to combine the use of sticks and carrots. Be that as it may, it is conspicuous that China's more cooperative, or even integrative, attitude in the Mekong region has primarily played out in areas other than (narrowly-defined) water resources management; the only such water resources management cooperation is China's signing of the Agreement on the Provision of Hydrological Information on the Lancang/Mekong River in 2002, under which China has provided the MRC with water-level data from its territory during each year's flood season (MRC, n.d.).[6]

Instead, much of China's integration strategy was initially centred on its economic engagement with and in the Greater Mekong Subregion (GMS). The GMS is a 1992 initiative of the Asian Development Bank (ADB) and its members include all six Mekong River riparian countries. The main objective of the mechanism is to develop GMS markets and facilitate the movement of goods and people across the Mekong region, thereby putting an emphasis on enhancing infrastructure linkages in particular (ADB 2011). Within the GMS, China quickly moved from beneficiary to benefactor (Lim 2008) and has therefore emerged as the "top trader, investor and donor" in the region (Chheang 2010: 362).

More precisely, Biba (2012) has pointed out that China's aggregate bilateral trade with the downstream Mekong countries (excluding Myanmar) grew more than 30-fold, from 1.6 billion to 53 billion USD in the period from 1990 to 2007 (IMF, n.d.). Using the same IMF database for more recent years shows that this trend has continued through the 2010s. In 2013, China's trade with the same countries amounted to 142.8 billion USD; in 2018, it reached 247.4 billion USD (IMF, n.d.). A similar pattern can also be revealed for China's foreign direct investment (FDI) in the Mekong region (again excluding Myanmar). While it had already increased more than six-fold, from nearly 93 million to over 580 million USD between 2003 and 2009 (Biba 2012), the 2017 figures for Thailand and Vietnam have both crossed the 1 billion USD threshold (National Bureau of Statistics of China 2018). Given that the Mekong region is included in the China-Indochina Peninsula Economic Corridor,

[6] It should also be mentioned that China has rejected (regularly) sharing hydrological data during the dry season, in spite of MRC requests (Biba 2018a). Also see Rampini (this volume) regarding how this issue is addressed in the Yarlung-Tsangpo-Brahmaputra basin.

one of six main corridors of China's omnipresent Belt and Road Initiative (BRI), China's economic interests in the region are almost certainly bound to only increase further.

One illustrative example of a specific area where China has been particularly interested in pushing forward (economic) integration is transport infrastructure. From early on, China has focused on developing the GMS North–South Corridor, which includes its provinces of Yunnan and Guangxi and stretches southward in two different directions to Bangkok and Hanoi, respectively. Beijing has been heavily involved in building roads to improve regional trade infrastructure (Lee 2015). As road infrastructure has now made extensive headway, China has recently sought to use the Lancang-Mekong Cooperation (LMC) mechanism (see below) to upgrade the railway network between China and mainland Southeast Asia (Lu 2016). In addition, China has also long sought to make greater use of the Mekong River itself as a transportation route. Together with Myanmar, Laos, and Thailand, China inked the Agreement on Commercial Navigation on the Mekong-Lancang River in 2000, which was meant to pave the way for the blasting of rocks and rapids in the Mekong (Schmeier 2009). This controversial plan was soon stopped but has more recently re-emerged (Southerland 2017).

Lastly, it is noteworthy that China's integration strategy is no longer focused only on development, but has developed a distinctive security component as well. After an incident in October 2011 in which two Chinese cargo ships were assaulted on the Mekong River by suspected Myanmar-based drug traffickers, so-called non-traditional security threats along the river grew in significance for China.[7] To ensure river trade security, Beijing immediately pushed for a joint law enforcement and security cooperation mechanism among China, Myanmar, Laos, and Thailand (Biba 2018b). Between December 2011 and December 2019, 89 joint river patrols were conducted under China's aegis (Xinhua 2019).

[7] In this case, non-traditional security threats include non-military elements such as smuggling, drug trafficking, and illegal migration. For a brief conceptual overview of the similarities and differences between traditional and non-traditional security threats, see Biba (2016a).

Containment Strategy

While China has long pursued a hegemonic strategy of mixing sticks (resource capture) with carrots (integration), it has also come to understand the value of constructing a narrative framing the country's behaviour in a positive light. Zeitoun and Warner (2006) regard this ideological compliance mechanism as the silver bullet. At least since the 2008 GMS summit, China has been anxious to sell its goals and actions in the Mekong region under the benign slogan of "common development and prosperity", thereby presenting a "win-win" situation for China and its neighbours (Wen 2008: online). Accordingly, China has continuously portrayed itself as a benevolent upstream country that "always pays great attention to the concerns and requirements of countries in the downstream reaches" (Yang 2019: online). In the past, this 'attention' was exclusively focused on touting China-sponsored economic growth as a panacea to downstream countries' problems (Biba 2019). Recently, China's official rhetoric has sought to add a 'green label' to its narrative, arguing that the country is interested in striking a balance between economic development and ecological conservation. In doing so, China apparently hopes to corroborate its framing of the Mekong as a "river of friendship, cooperation and prosperity" (Yang 2019: online).

China has also decided to supplement its ideological co-optation strategy with a normative thrust, primarily via the LMC. This agreement brings together all six Mekong River riparian countries and was officially launched in March 2016 during an inaugural leaders' meeting held on China's Hainan Island. Previously, China's premier Li Keqiang had pitched the idea of the new mechanism at the 2014 summit of the Association of Southeast Asian Nations in Myanmar, indicating that China has been in the driver's seat from the very outset (Biba 2018b).

The LMC is a body that reflects Chinese interests in the Mekong region in a number of ways. First, in terms of its agenda, the LMC is the first China-led multilateral initiative that officially includes sustainable water resources management on its agenda. This is very much in contrast to China's refusal to become a member of the MRC partly because of its sole focus on water resources management (Backer 2007) as well as China's reported blocking of water issues from the remit of the GMS (Hensengerth 2009). China has thus tried to learn from past mistakes that have led to serious accusations regarding its hydro-political cooperation record (see below). Meanwhile, water resources management is only one of five LMC priority areas—and certainly not the most important. China clearly aimed to establish a comprehensive cooperation mechanism that combines the agendas of the MRC and the GMS and also integrates new issues of concern to China, particularly cooperation on non-traditional security problems (Biba 2018b).

The second aspect in line with Chinese ideals is the LMC's institutional design. Unlike the MRC, for example, the LMC is based on a 'project-oriented model' and, significantly, not on codified rules and procedures of cooperation (Middleton and Allouche 2016). While all LMC members are officially said to be equal, the loose institutional framework creates opportunities for China, as the most powerful member state, to be 'more equal than the others'. In this regard, it also is quite revealing how China seems to envision cooperation on water resources through the LMC. Zhang Jiuhuan, vice chairman of the China Public Diplomacy Association (a non-profit organisation aimed at projecting a positive image of China to the world via experts, scholars, and celebrities), has remarked that downstream countries "concerned about the implications of the dam construction upstream… could raise the issue for consultation with countries upstream", implying a role for power relations (cited in Biba 2018b: 636).

Finally, membership is controlled by China as well. While China's LMC proposal was attractive enough to pull in the other five Mekong riparian countries, China has simultaneously managed to exclude external—that is, non-riparian—countries. This is in stark contrast to both the MRC, which has (at least until recently) been sponsored largely by a group of mostly Western donors, and the GMS, an ADB initiative in which Japan and the USA have the biggest say (Biba 2018b). As a result, the LMC can be interpreted as China's 'baby', increasing the country's own institutional capacity in the Mekong region while keeping external influence at bay. Moreover, due to its loose connection with China's BRI, the LMC potentially has huge financial resources at its disposal (Middleton and Allouche 2016). Against this backdrop, China is free to set its preferred agenda in the LMC and above all push forward its ambitious programme for regional infrastructure enhancement.

Reactions from Downstream and Beyond

How have its riparian neighbours, as well as external great powers interested in regional affairs, reacted to China's hegemonic behaviour?

As the framework designed by Zeitoun and Warner (2006) demonstrates, hegemonic control entails various forms of *inter*action between the hegemon and riparian neighbours. This, in turn, implies that other (riparian and external) countries' *re*action to the hegemon's behaviour may speak volumes in terms of how this behaviour is valued. For China's riparian neighbours, to begin with, it is crucial to distinguish between the responses from state and non-state actors. While the former have—maybe with the exception of Vietnam—not generally pointed a finger at China, at least not officially, the latter have repeatedly been at the forefront of openly criticising China's hydro-politics. How come?

For one thing, more broadly speaking, China's huge economic clout in the region has created dependencies for downstream neighbours. This is particularly true for the poorer Mekong countries such as Laos and Cambodia, but even for Vietnam, which is relatively more developed and not generally China-friendly (cf. Malczyk and Robinson 2019). Under such circumstances, it has been difficult for downstream country leaders to speak up against China, even if they want to. More specifically speaking, downstream governments have not actually been very opposed to China's upstream dam building. In fact, Mehtonen (2008: 161) has poignantly contended:

> It is often argued that China builds dams on the upstream Mekong despite the opposition of the downstream countries. Particularly the media have reported the situation as if the downstream nations wished China to stop the damming. However, the opposing voices represent mostly the civil society and the NGOs, and not the national governments. Actually, all of the Mekong countries are involved in the plans for regional power trade, in which the Yunnan province of China will be a net exporter.[8]

What is more, China has a noteworthy role to play in downstream dam building. As downstream governments have in recent years increasingly attempted to develop the river's hydropower potential and reap the resulting economic benefits, China has provided them with both a crucial precedent and major investments. According to some estimates, Chinese state-owned enterprises will implement up to 40% of the proposed mainstream and tributary hydropower schemes in the downstream Mekong

[8] Also see the chapter by Hennig and Magee, this volume.

River countries in the coming years (Hirsch 2011; International Rivers 2014).

The picture has meanwhile looked very different for various non-state actors, mainly local communities that make their living directly from the Mekong's resources as well as local, regional, and global environmental protection and sustainable development NGOs. They have continually challenged China's preferred discourse of joint development, as fittingly illustrated by the motto: 'I want to eat fish. I cannot eat electricity' (Earth Rights International, n.d.).

A major episode in non-state actors' struggle against Chinese dams happened during a drought in spring 2010, when water levels in the lower Mekong river sank dramatically. Many in downstream countries who saw their livelihoods threatened started to blame Chinese dams on the Mekong's upper reaches for storing water and thus causing, or at least contributing to, the extremely low water levels. Supported by regional and international activists as well as the media inside and outside the region, local communities propelled a weeklong series of statements, reports, and articles that were harshly critical of China's dam building activities and the opacity surrounding them. They demanded that China stop building dams on the Lancang and that it release data detailing dam operations as well as pre-damming hydrological records (Biba 2016b, 2018a).

Today, the anger of local communities towards China's practices continues unabated. The previously mentioned situation related to China's maintenance work at Jinghong Dam represents the most recent case in point. Using the *Bangkok Post* to trade barbs with a spokesperson for the Chinese embassy in Thailand, Niwat Roykaew, a longstanding and prominent local activist, urged China to be "sincere" (Roykaew 2019: online). After the Chinese spokesperson had argued that China was "promoting water resources cooperation in the Mekong for the benefit of the people in the region", Roykaew countered with reference to China's upstream dams: "Let's be clear. Local riparian communities are not beneficiaries of the Lancang cascade". As a consequence of the Lancang dams altering the Mekong's natural and seasonal flow, he said, "[w]e have seen our river banks and boats washed away. We have seen a major decline in our fisheries and river weed, which has impacted our ability to make a living" (Roykaew 2019: online).

In the light of this situation, it is noteworthy to at least briefly mention that external great powers, primarily the USA and Japan, have set up comprehensive sub-regional cooperation mechanisms with all the Mekong riparian countries except China as members. Back in 2009, the USA established the Lower Mekong Initiative and Japan founded the Japan-Mekong Summit. In August 2019, the two countries moreover announced the creation of a joint Mekong Power Partnership aimed at promoting economic growth through open and stable energy markets (US Embassy in Burma 2019). Those initiatives are generally welcomed by downstream countries and show that the USA and Japan are still seeking to rein in Chinese influence in the Mekong region.

Conclusion

Applying the hydro-hegemony framework as introduced by Zeitoun and Warner (2006), this chapter has demonstrated that China's behaviour in the Mekong region has pursued the whole spectrum of control strategies available to hydro-hegemons. On the one hand, this is certainly in accordance with China's extremely predominant power resources in the Mekong region, which give the country enormous leeway in its actions. On the other hand, the mixed strategy also makes it difficult to rule out whether China is to be seen as a good or bad hydro-hegemon. Rather, it shows that China has been struggling with itself. The hegemon has been reluctant to share the Mekong River's water resources equally with its riparian neighbours. In fact, China has sought to capture and utilise these resources primarily to its own benefit and has paid little attention to potential negative impacts outside its borders. Yet China has been the key driver for regional integration mechanisms and has promoted cooperation on topics beyond narrowly-defined water resources management.

In this regard, it may well be argued that China has long instrumentalised its economic footprint to foster socio-economic progress in many places across the Mekong region (see Lee 2015). Progress has manifested in huge increases in average per capita incomes throughout the region, greatly reduced poverty levels, and sinking infant mortality rates, as well as markedly increased life expectancy, school enrolment, and literacy rates (ADB 2012). At the same time, though, it must be emphasised that China has not acted out of altruism. Instead, China's strong economic involvement in the Mekong region has followed an obvious domestic rationale emphasising economic development on its south-western periphery. The positive impacts that neighbours have experienced have ultimately not been much more than a welcome by-product (see Hensengerth 2009).

What is more, there also are many in the Mekong region who have not benefitted from China's increased engagement. Here, it is useful to assess the varied reactions to China's behaviour. Downstream governments—with the exception of Vietnam, at least to some degree—have quite successfully been co-opted by China and Chinese interests. The fact that all downstream countries nowadays conduct dam building activities along the Mekong River's mainstream and tributaries testifies to this. As opposed to governments, a great number of non-state actors across the region have contested how China has pursued its hydro-hegemony. It is interesting to note that in fighting back against China, non-state actors have particularly sought to construct knowledge, promote discursive alternatives, and raise awareness across and beyond the region. To use jargon in line with Zeitoun and Warner's hydro-hegemony framework, the non-state actors have sought to contest China's ideational hegemony. However, given limited governmental support at home, these non-state actors have not been able to effectively challenge China's behaviour, at least not over extended periods of time. While external great powers such as the USA and Japan have been sympathetic to these non-state actors' concerns, American and Japanese initiatives in the region to date have similarly lacked the resources needed to seriously challenge China's agenda—let alone its predominant position.

In sum, then, China's hydro-hegemony in the Mekong region can be characterised neither as fully positive leadership nor as completely negative domination. Instead, China's performance has exhibited features of what one would expect from both a good and a bad hydro-hegemon, though it is probably fair to say that China's hydro-hegemony has been more harmful than helpful. China therefore still has a long way to go, should the country become interested in being a more benevolent leader in the Mekong region. This is particularly true when it comes to sharing the Mekong basin's resources more equitably and giving other actors a bigger say with regard to the river's future trajectory.

Post-scriptum

During the production of this book, much has happened in the Mekong region that holds relevance to the arguments presented in this chapter. Parts of the Mekong region went through the "worst drought ever" in 2019–2020 (Khoa 2020). The region is also increasingly becoming a new battleground for Sino-US strategic competition (Siow 2020). Seeking to counter Chinese influence in mainland Southeast Asia more effectively, the US has stepped up its presence, particularly by establishing the Mekong-US Partnership in September 2020 (US Department of State 2020). All of these developments have led China to decide to share year-round hydrological data with downstream Mekong countries (Reuters 2020). The fact that China has finally agreed to this long-standing MRC request is a positive sign and signals the potential for greater transparency and collaboration among riparian stakeholders. However, as before, this has been a purely reactive Chinese measure precipitated by crisis and criticism. A true leader would stand out for much more (pro)active cooperation.

References

Asian Development Bank (ADB). (2011). *The Greater Mekong Subregion economic cooperation program strategic framework 2012–2022*. Manila: ADB.

Asian Development Bank (ADB). (2012). *The Greater Mekong Subregion at 20: Progress and prospects*. Manila: ADB.

Backer, E. (2007). The Mekong River Commission: Does it work, and how does the Mekong basin's geography influence its effectiveness? *Südostasien Aktuell, 26*(4), 31–55.

Biba, S. (2012). China's continuous dam-building on the Mekong River. *Journal of Contemporary Asia, 42*(4), 603–628.

Biba, S. (2016a). New concerns, more cooperation? How non-traditional security issues affect Sino-Indian relations. *Journal of Current Chinese Affairs, 45*(3), 3–30.

Biba, S. (2016b). From securitization moves to positive outcomes: The case of the spring 2010 Mekong crisis. *Security Dialogue, 47*(5), 420–439.

Biba, S. (2018a). *China's hydro-politics in the Mekong: Conflict and cooperation in light of securitization theory*. London: Routledge.

Biba, S. (2018b). China's 'old' and 'new' Mekong River politics: The Lancang-Mekong Cooperation from a comparative benefit-sharing perspective. *Water International, 43*(5), 622–641.

Biba, S. (2019, August 30). China's hegemonic choice in the Mekong region. *East Asia Forum*. https://www.eastasiaforum.org/?p=202337. Accessed 3 June 2020.

Cascao, A. (2008). Ethiopia—Challenges to Egyptian hegemony in the Nile Basin. *Water Policy, 10*(S2), 13–28.

Cascao, A., & Zeitoun, M. (2010). Power, hegemony and critical hydropolitics. In A. Earle, A. Jägerskog, & J. Öjendal (Eds.), *Transboundary water management: Principles and practice* (pp. 27–42). London: Earthscan.

Central Intelligence Agency (CIA). (2019). *The world factbook*. https://www.cia.gov/library/publications/the-world-factbook/. Accessed 3 June 2020.

Chellaney, B. (2011). *Water: Asia's new battleground*. Washington: Georgetown University Press.

Chellaney, B. (2013, February 7). China's hydro-hegemony. *New York Times*. https://www.nytimes.com/2013/02/08/opinion/global/chinas-hydro-hegemony.html. Accessed 3 June 2020.

Chheang, V. (2010). Environmental and economic cooperation in the Mekong region. *Asia Europe Journal, 8*(3), 359–368.

Dahl, R. A. (1965). *A preface to democratic theory*. Boulder: University of Chicago Press.

De Stefano, L., Edwards, P., de Silva, L., & Wolf, A. (2010). Tracking cooperation and conflict in international basins: Historic and recent trends. *Water Policy, 12*(6), 871–884.

Earth Rights International. (n.d.). *I want to eat fish. I cannot eat electricity: Public participation in Mekong basin development*. https://earthrights.org/wp-content/uploads/publications/I-Want-To-Eat-Fish.pdf. Accessed 3 June 2020.

Eyler, B., & Weatherby, C. (2020, April 13). New evidence: How China turned off the tap on the Mekong River. *Stimson*. https://www.stimson.org/2020/new-evidence-how-china-turned-off-the-mekong-tap/. Accessed 3 June 2020.

Hensengerth, O. (2009). Transboundary river cooperation and the regional public good: The case of the Mekong river. *Contemporary Southeast Asia, 31*(2), 326–349.

Hirsch, P. (2011). China and the cascading geopolitics of lower Mekong dams. *The Asia-Pacific Journal, 9*(2). https://www.japanfocus.org/-Philip-Hirsch/3529.

Ho, S. (2016). 'Big brother, little brothers': Comparing China's and India's transboundary river policies. *Water Policy, 18*(S1), 32–49.

Hutt, D. (2019, October 16). Water war risk rising on the Mekong. *Asia Times*. https://www.asiatimes.com/2019/10/article/water-war-risk-rising-on-the-mekong/. Accessed 3 June 2020.

International Monetary Fund (IMF). (n.d.). *Direction of trade statistics*. https://data.imf.org/?sk=9D6028D4-F14A-464C-A2F2-59B2CD424B85. Accessed 3 June 2020.

International Rivers. (2014). *China overseas dams list*. https://www.internationalrivers.org/resources/china-overseas-dams-list-3611. Accessed 3 June 2020.

International Rivers. (2019). *Understanding the impacts of China's upper Mekong dams*. https://www.internationalrivers.org/resources/8477. Accessed 3 June 2020.

Ketelsen, T., Räsänen, T., & Sawdon, J. (2020, May 29). Opinion—Did China turn off the Mekong tap? Why data matters for cooperation. *The*

Third Pole. https://www.thethirdpole.net/2020/05/29/opinion-did-china-close-mekong-tap-why-data-matters-for-cooperation/. Accessed 3 June 2020.

Khoa, H. (2020, March 21). Mekong Delta hit by worst drought ever. *VN Express*. https://e.vnexpress.net/photo/news/mekong-delta-hit-by-worst-drought-ever-4071241.html. Accessed 26 October 2020.

Lee, S. (2015). Benefit sharing in the Mekong River basin. *Water International, 40*(1), 139–152.

Lim, T. (2008). *China's active role in the Greater Mekong Sub-Region: A "win-win" outcome?* (Background brief no. 39). Singapore: East Asia Institute.

Lu, G. (2016). *China seeks to improve Mekong sub-regional cooperation: Causes and policies* (Policy Report). Singapore: RSIS.

Lukes, S. (2005). *Power: A radical view*. Hampshire: Palgrave Macmillan.

Malczyk, B., & Robinson, T. (2019, July 10). Vietnam needs to break its addiction to Chinese coal. *The Diplomat*. https://thediplomat.com/2019/07/vietnam-needs-to-break-its-addiction-to-chinese-coal/. Accessed 3 June 2020.

McCracken, M., & Wolf, A. (2019). Updating the register of international river basins of the world. *International Journal of Water Resources Development, 35*(5), 732–782.

Mehtonen, K. (2008). Do the downstream countries oppose the upstream dams? In M. Kummu, M. Keskinen, & O. Varis (Eds.), *Modern myths of the Mekong: A critical review of water and development concepts, principles and policies* (pp. 161–172). Helsinki: Helsinki University of Technology.

Mekong River Commission (MRC). (n.d.). *Upstream partners*. https://www.mrcmekong.org/about-mrc/upstream-partners/. Accessed 3 June 2020.

Menga, F. (2016). Reconceptualizing hegemony: The circle of hydro-hegemony. *Water Policy, 18*(2), 401–418.

Middleton, C., & Allouche, J. (2016). Watershed or powershed? Critical hydropolitics, China and the 'Lancang-Mekong Cooperation framework.' *International Spectator, 51*(3), 100–117.

National Bureau of Statistics of China. (2018). *China statistical yearbook, 2018*. https://www.stats.gov.cn/tjsj/ndsj/2018/indexeh.htm. Accessed 3 June 2020.

Nickum, J. (2008). The upstream superpower: China's international rivers. In O. Varis, C. Tortajada, & A. Biswas (Eds.), *Management of transboundary rivers and lakes* (pp. 227–244). Berlin: Springer-Verlag.

Reuters. (2020, October 21). China commits to share year-round water data with Mekong River Commission. *Reuters*. https://www.reuters.com/article/us-mekong-river/china-signs-pact-to-share-year-round-water-data-with-mekong-river-commission-idUSKBN277135. Accessed 26 October 2020.

Roykaew, N. (2019, July 17). China must be sincere on Mekong. *Bangkok Post*. https://www.bangkokpost.com/opinion/opinion/1713756/china-must-be-sincere-on-mekong. Accessed 3 June 2020.

Schmeier, S. (2009). Regional cooperation efforts in the Mekong River basin: Mitigating river-related security threats and promoting regional development. *Austrian Journal for Southeast Asian Studies, 2*(2), 28–52.

Siow, M. (2020, August 3). The next US-China battleground: Chinese dams on the Mekong River? *South China Morning Post*. https://www.scmp.com/print/week-asia/politics/article/3095581/next-us-china-battleground-chinese-dams-mekong-river. Accessed 26 October 2020.

Southerland, D. (2017, January 27). The Mekong part 4: Blasting the rapids in Thailand. *Radio Free Asia*. https://www.rfa.org/english/thailand-mekong-01272017104723.html. Accessed 3 June 2020.

Strange, S. (1987). The persistent myth of lost hegemony. *International Organization, 41*(4), 551–574.

US Department of State. (2020). Launch of the Mekong-U.S. Partnership: Expanding U.S. engagement with the Mekong Region. Fact sheet. https://www.state.gov/launch-of-the-mekong-u-s-partnership-expanding-u-s-engagement-with-the-mekong-region/. Accessed 26 October 2020.

US Embassy in Burma. (2019). *Joint statement on the Japan-United States Mekong Power Partnership (JUMPP)*. https://mm.usembassy.gov/joint-statement-on-the-japan-united-states-mekong-power-partnership-jumpp/. Accessed 3 June 2020.

Warner, J., Mirumachi, N., Farnum, R., Grandi, M., Menga, F., & Zeitoun, M. (2017). Transboundary 'hydro-hegemony': 10 years later. *WIREs Water, 4*(6), e1242.

Wen, J. (2008, March 31). *Build a bond of cooperation and a common homeland*. Speech at the 3rd GMS summit, Vientiane, Laos. https://www.china-embassy.org/eng/zt/768675/t476515.htm. Accessed 3 June 2020.

Xinhua. (2019, December 28). Joint patrol concludes on Mekong River. *Xinhua*. https://www.xinhuanet.com/english/2019-12/28/c_138663125.htm. Accessed 3 June 2020.

Yang, Y. (2019, July 12). False report undermines Mekong cooperation. *Bangkok Post*. https://www.bangkokpost.com/opinion/opinion/1711051/false-report-undermines-mekong-cooperation. Accessed 3 June 2020.

Young, O. (1991). Political leadership and regime formation: On the development of institutions in international society. *International Organization, 45*(3), 281–308.

Zeitoun, M., & Warner, J. (2006). Hydro-hegemony—A framework for analysis of trans-boundary water conflicts. *Water Policy, 8*(5), 435–460.

CHAPTER 12

Hydropower and Sino-Indian Hydropolitics Along the Yarlung-Tsangpo-Brahmaputra

Costanza Rampini

Introduction

The Yarlung-Tsangpo-Brahmaputra River (YTB) is one of the world's largest river systems (Pahuja and Goswami 2006). Originating in the Angsi Glacier in Tibet, the basin spreads over more than 500,000 km^2 of land in China, India, Bhutan, and Bangladesh, though over 80% of it lies in China and India (Mohammed et. al 2017). In India, the YTB drains the north-eastern region, with over 85% of the sub-basin shared between the states of Arunachal Pradesh and Assam. After cutting through the plains of Assam, the YTB enters Bangladesh, where it meets the Padma and Meghna Rivers before flowing into the Bay of Bengal (See Fig. 12.1) (Bandyopadhyay et al. 2016). Approximately 130 million people live within the YTB river basin, many of whom are rural, poor, and rely on the river for their sustenance and livelihoods (Jiang et al. 2017). Indeed, in North-east India, the YTB is often referred to as the 'lifeline' of the

C. Rampini (✉)
San José State University, San José, CA, USA
e-mail: costanza.rampini@sjsu.edu

J.-F. Rousseau and S. Habich-Sobiegalla (eds.), *The Political Economy of Hydropower in Southwest China and Beyond*, International Political Economy Series, https://doi.org/10.1007/978-3-030-59361-2_12

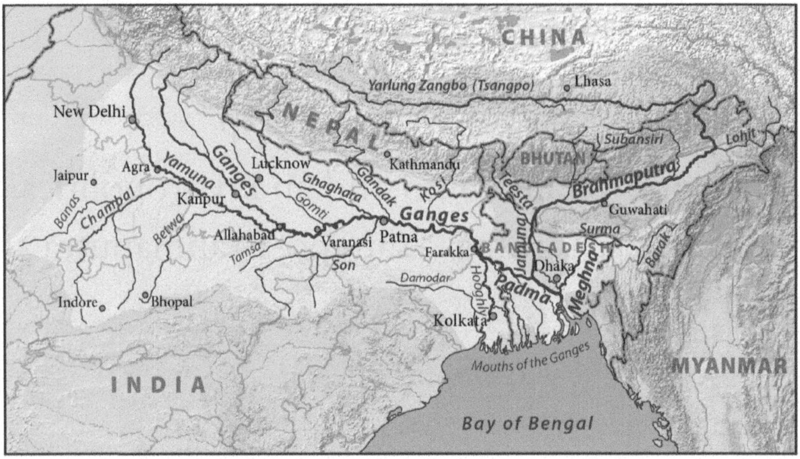

Fig. 12.1 The Yarlung-Tsangpo-Brahmaputra river system (*Source* Pfly [2011], licensed under the Creative Commons Attribution-Share Alike 3.0 Unported license. https://commons.wikimedia.org/wiki/File:Ganges-Brahmaputra-Meghna_basins.jpg. Accessed 18 June 2020)

region (Vagholikar and Das 2010). As the YTB rapidly descends from the Himalayan Mountains to the plains of Assam, it traverses steep slopes and gathers tremendous energy, which gets dissipated in the form of intense summer floods, especially in India and Bangladesh (Bora 2004; Bandyopadhyay et al. 2016).

The energy that the YTB gains throughout its course also puts the river at the centre of China's and India's recent renewable energy development strategies. Both countries are mobilising their scientific and engineering capacities to dam their respective stretches of the river and harness its hydropower potential. China has already completed the Zangmu Dam along the main stem of the Yarlung Tsangpo, and rumours of large-scale water diversion schemes targeting the river have emerged. Similarly, the Indian central government has expedited the clearance of big dams along the Indian stretches of the YTB and its tributaries, partly in an effort to counter Chinese hydro-ambitions. As both countries speed up efforts to develop the hydropower potential of the YTB, and in the absence of a formal water agreement between China and India, many have raised concerns about the intensification of Sino-Indian tensions. This situation is particularly worrisome, given that the YTB crosses an already disputed border[1] between the two countries, and hydropower development along its course threatens to compound long-standing tensions over Tibet and China's growing regional influence.

This chapter aims to shed light on the relationship between China and India over this shared river and attenuate fears of an imminent 'water war' between Asia's superpowers. It begins by discussing the countries' respective plans to develop the YTB and then explains how Sino-Indian hydropolitics along the river are embedded in a larger set of tensions between the two countries. Next, the chapter highlights the extent to which India and China already cooperate over the YTB and asks whether climate change impacts on the river might precipitate further collaboration. Finally, the chapter concludes by reflecting on the role of Sino-Indian hydropolitics in shaping the future of the Himalayan region and its mighty rivers.

Chinese and Indian Hydropower Ambitions Along the YTB

Despite the YTB's immense hydropower potential—the river alone possesses nearly 40% of India's entire assessed hydropower potential—until recently, it remained one of the last great undammed rivers in the world (CEA 2015). Since the early 2000s, however, both China and India have begun building dams along its course in an effort to meet increasing energy demands, and in response to mounting pressures to reduce their greenhouse gas emissions.

China has constructed several dams along tributaries of the YTB, such as the Pangduo and Zhikong Dams on the Lhasa River. In 2014, it completed the Zangmu Dam (510 MW), its first dam along the YTB main stem (Alam et al. 2016). Additionally, China's Twelfth Five Year Plan (2011–2015) outlined plans for three new dams along the river that are now under construction—one in Dagu (640 MW), another at Jiacha (320 MW), and a third of unconfirmed capacity at Jiexu (Krishnan 2013). While the Thirteenth Five Year Plan (2016–2020) promises 74.6

[1] China and India have disputed borders along three fronts: the western front in the Kashmir and Aksai Chin region, the middle front along the borders of Himachal Pradesh and Uttaranchal, and the eastern front, where China claims the state of Arunachal Pradesh (Panda 2017).

billion U.S. dollars (USD) of investments in new installed hydropower capacity (Koleski 2017), it mentions no specific projects and Chinese media have reported that the government will first focus on building new dams in Tibet that are not along the YTB (Walker 2016). Still, as China completes construction of these dams, its hydropower efforts are bound to shift towards the YTB, especially towards a site known as the 'Great Bend' where the YTB drops 1400 m just before crossing into India (He 2011). The Chinese government has assured India that, because of their small scale, dams along the Chinese stretch of YTB would have no downstream transboundary impacts. Indian officials, however, continue to voice concerns over China's lack of transparency about its schemes for this shared river (Tandon 2019). Perhaps even more concerning to Indian officials than Chinese dams along the YTB is China's controversial multi-billion-USD plan to divert water from its southern regions to its more arid regions, a project known as the South-North Water Transfer. Segments of this diversion scheme have been completed or are under construction, but the western route, which included schemes to divert the flows of the YTB, was seemingly scrapped due to debates over its technical and financial viability (Fan 2018).

Rather than adopting a forthright attitude with the Chinese government over the issue of hydropower development and water diversion schemes along the YTB, the Indian government has instead expedited the approval of big dams along its own stretches of the river. As of 2012, the state government of Arunachal Pradesh had cleared 140 new dam projects for construction along the YTB and its tributaries, for a total installed capacity of over 40 GW (Vagholikar and Das 2010). Besides increasing its energy supply, India is also using hydropower development to establish first-user rights to the flows of the YTB (Kurian 2013, Bhattacharya 2018). In other words, despite the lack of any water treaty between China and India upholding these specific rights, India is building dams to appeal to the well-known water doctrine of prior-appropriation, which holds that the first user of water has the right to continue using that same quantity of water, and subsequent users cannot interfere with their ability to do so (Lazerwitz 1993). Any future Chinese dam or diversion scheme along the YTB could affect the capacity of existing Indian dams downstream to generate hydropower—that is, unless China respects the prior-appropriation rights of pre-existing Indian dams.

Overall, China's and India's respective interests in developing the YTB have led to speculations about the potential of a 'water war' between the two countries, as they each race to tap the hydropower potential of the river. Specifically, many are concerned that their respective hydro-ambitions will exacerbate other long-standing Sino-Indian conflicts in the region (Chellaney 2011). The following section further explores the ways in which Sino-Indian hydropolitics along the YTB are complicated by existing conflicts between the two countries, both in the immediate region and beyond. It begins by explaining the Sino-Indian border dispute along the McMahon Line and tensions over Tibet's independence. It then explores the evolution of Sino-Indian relations since the election of Narendra Modi as Indian Prime Minister in 2014.

YTB Hydropolitics in the Sino-Indian Political Context

The Eastern Border Dispute and Tibet

There are 310 international rivers in the world, each with their own set of challenges when it comes to the management of their transboundary flows (McCracken and Wolf 2019). In the case of the YTB, water diplomacy between China and India is complicated by the fact that the river crosses one of the disputed boundaries between the two countries, known as the McMahon Line. The McMahon Line, which separates the Indian state of Arunachal Pradesh from Tibet, was negotiated in 1914 by representatives of the new Republic of China, the Tibetan government, and the British government. India and the international community continue to recognise it as the legal border between North-east India and today's Tibet Autonomous Region of China (Panda 2017). However, since gaining control over Tibet in the mid-twentieth century, China has contested the border, arguing that Tibet was not an independent state and could not sign territorial treaties (Panda 2017). This has led both China and India to establish a permanent military presence on their respective sides of the contested line and, in 1962, the border became the site of the last Sino-Indian War. Formal border talks between China and India started in 1981 and are now entering their 22nd round, but they have yielded no accord. Indeed, resolving the McMahon border dispute would require that China and India first reach an agreement over the Tibetan question.

The issue of Tibet's independence has been a sore topic between India and China since 1951, when China established de facto control over the Tibetan territory (Panda 2017). Then in April 1959, days before the Tibetan revolt against Chinese troops in Lhasa, the Dalai Lama fled the Tibetan capital and travelled on foot to Arunachal Pradesh, where he was granted asylum by the Indian government. Today, India is home to the Dalai Lama, the Tibetan-government-in-exile, and over 94,000 Tibetans—the largest settlement outside of Tibet (Central Tibetan Administration 2019). While Indian officials often cited cultural kinship as a reason for supporting Tibet, they were also wary of the fact that, if China asserted control over Tibet, the territorial buffer between the two countries would be lost and this could re-open border negotiations along the McMahon Line (Garver 2001). Additionally, it meant that China would control vast natural resources in the Tibetan Plateau, including the headwaters of all of Asia's main rivers (Garver 2001). Eventually, in October 1962, territorial disputes, growing tensions over the Tibetan issue, and Cold War rivalries culminated in the last Sino-Indian war, which was partially fought along the disputed McMahon Line (Panda 2017).

India lost the war but maintained sovereignty over Arunachal Pradesh. Since then, and especially since the beginning of the border talks in 1981, tensions between China and India over their disputed eastern border and the Tibetan issue have de-escalated, while other forms of bilateral relations have improved (Singh 2018). However, these unresolved conflicts remain the main stumbling blocks to any form of true Sino-Indian collaboration or treaty over the YTB, because they have led to the securitisation of its flows (R. Bhattacharjee, personal communication, 28 June 2019). As a result of the border issue, the waters of the YTB have become a matter of national security for both India and China, meaning that hydrological data about the basin is often classified as pertaining to national defence and not available to researchers (Bandyopadhyay and Ghosh 2009, Bhambri and Bolch 2009). Similarly, due to restricted access to stretches of the YTB in both countries, the basin is considerably under-researched compared to other Asian rivers, and the data sets that exist have large gaps, often coinciding with key geopolitical events such as the 1962 war (Ray et al. 2015, Rampini 2016).

Modi's Stance on China and Border Disputes

Since the mid-1980s, India and China have made no significant progress in resolving their eastern border dispute, leaving little hope for Sino-Indian cooperation over the YTB. However, the election of Narendra

Modi as the 14th Prime Minister of India in May 2014 seemed to signal a change in Sino-Indian relationships and India's approach to boundary disputes. During his time as Chief Minister of Gujarat (2001–2014), Modi developed close ties with China and Chinese investors, and many hoped that his election would bring Asia's superpowers closer (R. Bhattacharjee, personal communication, 28 June 2019). At the same time, during his campaign, Modi promised to take a tough line on the border issue (Burke 2014). Both factors suggested that, more than any of his predecessors, Modi had the necessary attributes to initiate a bilateral water dialogue between India and China over the shared YTB, and perhaps even bring up the subject of hydropower development along its flows.

However, despite an increase in bilateral trade between two countries, many argue that Sino-Indian tensions have slightly risen since Modi's election. In 2016, China opposed India's bid to join the Nuclear Supplier's Group, and it put a technical hold on India's request to add three Pakistani-based individuals to the United Nations Security Council list of global terrorists (Singh 2018). Additionally, Modi's government has formally objected to China's One Belt One Road (OBOR) initiative, which aims to revitalise its relationships with countries along the ancient Silk Road via infrastructure development and trade agreements. India has long been wary of China's growing influence in Asia, which it has not been able to replicate via its own initiatives such as Modi's Act East policy and Link West strategy (Panda 2017). Moreover, India specifically opposes the China-Pakistan Economic Corridor component of the OBOR initiative because it includes projects in the Kashmir region, which is disputed by Pakistan and India (MEA 2017).

Amongst the recent flare-ups between China and India since Modi took office, the 2017 Doklam standoff has had by far the greatest impact on bilateral relations over the YTB. The Doklam Plateau is a territory disputed by China and Bhutan that lies near the North-east region of India and within the YTB river basin. In June 2017, Chinese troops moved into the plateau to presumably extend an existing road, though many interpreted the military incursion as an attempt to weaken India's long-standing relationship with Bhutan (Panda and Baruah 2019). In response, India took the international community by surprise and sent its troops to Doklam to face-off with Chinese troops on Bhutanese territory (Blank 2017). The episode initially raised fears of a wider escalation between Asia's superpowers, but the standoff was resolved two months later, as both nations agreed to withdraw their troops from the plateau

(Blank 2017). Less than a year later, Prime Minister Modi and President Xi Jinping held the first-ever bilateral informal summit in Wuhan, in part to restore positive Sino-Indian relations in the aftermath of Doklam (Godbole 2018). While the 2018 Wuhan Summit re-established the *status quo* and the countries' tacit understanding not to let boundary issues complicate other bilateral relationships (Panda and Baruah 2019), the Doklam episode revived Sino-Indian tensions over their disputed borders and the shared YTB river. Indeed, during and following the standoff, China stopped sharing hydrological data with India from its stations along the YTB, which it had been providing every summer according to a Memorandum of Understanding (MoU) signed between the two countries in 2002 (see next section).

Overall, despite initial optimism about Modi's relationship with China and his tough stance on border issues, under his term Sino-Indian relationships have seesawed, and the two countries are no closer to coming to a resolution over their borders. Additionally, despite portraying himself as strongman of Indian politics, Modi has never directly brought up hydropower development along the YTB with his Chinese counterpart, and his heavy-handed reaction to the Doklam episode threatened to put an end to the only form of Sino-India water diplomacy that is currently taking place along this shared river.

Sino-Indian Hydro-Diplomacy Along the YTB

Despite the potential for transboundary rivers to stoke conflict between riparian countries, historical examples of positive water diplomacy and cooperation outnumber international water conflicts (Wolf 2003). This section highlights two examples of Sino-Indian cooperation along the YTB and emphasises the need for more such efforts, given the threat that climate change poses to the flows of the YTB and the communities that rely on them.

Data Sharing and the Joint Expert Level Mechanism on Trans-Border Rivers

As of today, there is no international water treaty governing the YTB or guiding the countries in the river basin as they compete over its uses and development. Additionally, China and India both refused to sign the 1997 United Nations Convention on the Law of Non-Navigational Uses of International Watercourses, which provides a framework to ensure the equitable and reasonable utilisation of transboundary rivers (McCaffrey 2008, see also Biba in this volume). In sum, there are no international legal instruments requiring the countries of the YTB basin to cooperate regarding this shared resource (Bandyopadhyay et al. 2016). Bangladesh, as the lowest riparian country in the basin, is in an especially precarious position. The legal limbo also casts doubt on whether China would even recognise India's first-user rights along the YTB.

Nonetheless, while there exists no legal framework to negotiate Sino-Indian relations over the YTB, the two countries are engaged in a tenuous collaboration over the issue of flood forecasting along the river. Thanks to a MoU signed in 2002, the Indian Central Water Commission has been receiving flood season hydrological data from Chinese authorities twice a day from 1 June to 1 October. As part of the MoU, and despite the fact that data from the YTB is often classified as pertaining to national defence, the Chinese government has agreed to provide water level, discharge, and rainfall data from three stations along the Chinese stretch of the Yarlung Tsangpo in Nugesha, Yangcun, and Nuxia (NPCMT 2011, MOWR 2017). The data are then disseminated by the Indian Central Water Commission to other agencies, such as the Assam State Disaster Management Authority, to help improve flood preparedness and response (M. Irshad, personal communication, 23 June 2019). This is significant because during the summer months, the waters of the YTB cause devastating floods in North-east India, particularly in the state of Assam, where 40% of the land is vulnerable to riverine flooding (NRSC 2016). The provision of hydrological data from the upper reaches of the YTB basin is thus an important step in improving early warning systems and emergency preparedness protocols in the region.

Starting in 2006, the Chinese and Indian governments also set up a Joint Expert Level Mechanism (ELM) on Trans-border Rivers. As part of this ELM, expert groups from both countries meet on a yearly basis to discuss flood emergency management and the governance of their shared rivers, including the YTB (MOWR 2017). The expert groups are led by the Commissioner of the Indian Ministry of Water Resources, and by the Director of the International Economic and Technical Cooperation and Exchange Centre of the Chinese Ministry of Water Resources, respectively (NPCMT 2011). While these yearly bilateral meetings serve to create the image of a "cordial and friendly atmosphere" between the two countries over the issue of transboundary rivers (Embassy of the PRC in India

2019), they have yielded little results in terms of actual collaboration and shared governance.

The most notable development occurred in December 2013, when the Chinese government agreed to extend the flood data sharing period with India by two weeks, starting on May 15 rather than June 1 (MEA 2013). Though heavy floods along the YTB generally occur between June and August as a result of the monsoon rains, the first pulse of floods can arrive as early as late April. Moreover, extending the data sharing period will become increasingly necessary as a result of climate change impacts in the basin (J. Gupta, personal communication, 27 June 2019). Unfortunately, in the aftermath of the Doklam standoff, Chinese officials stopped sharing hydrological data from the YTB with India for the remainder of the 2017 monsoon season, and no joint ELM meeting over transboundary rivers was held that year. Chinese officials explained that they were unable to share flood data due to upgrades to their stations along the river (PTI 2018), and data sharing has since resumed. Nonetheless, the episode served as a stark reminder that Sino-Indian cooperation along the YTB is inextricably linked to Sino-Indian relations along their disputed eastern border and in the broader Asian region.

The Brahmaputra Dialogue

Given the lack of a formal water treaty to help govern the YTB, some have suggested the need for a transnational autonomous river basin organisation, similar to the Mekong River Commission, that could help foster the integrated management of the river and avoid the competitive exploitation of its flows by the basin countries (Hilton 2014, Bandyopadhyay et al. 2016). The Brahmaputra Dialogue, which began in 2013, is a notable attempt to foster this type of transnational water diplomacy along the YTB.

Also known as the *Transboundary Policy Dialogue for Improved Water Governance in the Brahmaputra River*, the Brahmaputra Dialogue was initiated by the South Asia Consortium for Interdisciplinary Water Resources Studies as a Track 3 dialogue involving relevant civil society stakeholders from the Brahmaputra basin countries.[2] The dialogue began with two main objectives: (1) initiating a shared data collection effort to

[2]This refers to the different tracks of diplomacy. *Track 1* describes diplomatic initiatives involving official actors, such as heads of state, diplomats, and high-ranking government

identify key concerns and assess the regional knowledge base in the YTB and (2) identifying key stakeholders in the basin and fostering a multilateral dialogue amongst them (M. D. Surie, personal communication, 24 July 2019). In its first phase (2013–2015), the dialogue was supported by seed funding from the Asia Foundation in Delhi, and only included stakeholders from India and Bangladesh (M. D. Surie, personal communication, 24 July 2019). Dr. Anamika Barua, one of the founders and current organiser of the dialogue, explained that they initially doubted Chinese stakeholders would participate in the forum, so they prioritised inviting Indian and Bangladeshi actors. However, by the end of the first phase, they decided to extend the invitation to Chinese scientists from Yunnan University, who eagerly accepted to participate in future workshops (A. Barua, personal communication, 20 June 2019). At the same time, the World Bank offered to fund the next phases of the dialogue, while also expressing its desire to see the forum move from a bilateral to a multilateral approach (Barua 2018a).

As a result, the second (2014–2015) and third phase (2015–2017) of the dialogue involved Bhutanese and Chinese stakeholders, and workshops where held in all four basin countries. Chinese participants included researchers from Yunnan University, Shanghai Institutes for International Studies, and Fudan University, amongst others (Barua 2018b; A. Barua, personal communication, 20 June 2019). The dialogue also shifted from a Track 3 to a Track 2–1.5 process, including not only academics and civil society actors, but also researchers working for government institutions, as well as former and current government officials (Barua 2018a). The dialogue is now entering its fourth phase, and the World Bank has signalled its willingness to continue funding this multilateral effort (A. Barua, personal communication, 20 June 2019).

Still, the dialogue needs to surmount several challenges in order for it to be effective at promoting the integrated management of the YTB. First, the dialogue has no real capacity or formal mechanism to affect water policy in the basin (Barua 2018a). Additionally, it is difficult to quantify its impacts, which limits the organisers' capacity to find donors willing to support these transnational multi-stakeholder workshops (Barua

officials. *Track 2* refers to informal diplomatic interventions that include retired government officials, regional and local leaders, academics, and influential non-governmental actors. Track 3 diplomacy, also known as 'people-to-people' diplomacy, takes place at the grass-roots level and does not involve official actors (Bohmelt 2010; Wasike et al. 2016).

2018a; A. Barua, personal communication, 20 June 2019). And while the dialogue has managed to successfully include Chinese academics, organisers have been unable to identify Chinese civil society organisations working on the YTB river (J. Gupta, personal communication, 27 June 2019). This is due to the fact that the Tibetan reaches of the YTB are sparsely populated, and that Chinese civil society organisations are restricted in their capacity to work in this remote and political sensitive region (Liu 2014). Additionally, no Tibet-based researcher has been invited to participate in the forum thus far (A. Barua, personal communication, 20 June 2019). Finally, while the topic of hydropower development has come up during the dialogue workshops (Barua 2018b), it is not at the core of these efforts. This is in part due to Sino-Indian tensions over the issue, but also because of frictions between the Indian government and Indian civil society members over dam-building efforts along the YTB. This friction also explains why, as the dialogue moves increasingly towards a Track 2–1.5 process with more government officials' involvement, it has become harder to bring Indian civil society actors to the table (M. D. Surie, personal communication, 24 July 2019).

Overall, the Brahmaputra Dialogue is the first water diplomacy platform which focuses exclusively on the YTB river basin and has successfully managed to include stakeholders from all four countries in the watershed. Other multilateral forums have attempted a similar task in the past, but they were often unable to get participation from Chinese stakeholders and they were not limited to the YTB river alone (Barua 2018a). The dialogue is still in its early stages and it is too soon to evaluate whether it will be able to foster the integrated management of the YTB river between China, Indian, Bhutan, and Bangladesh. Still, Dr. Anamika Barua, a current organiser of the dialogue for the World Bank, argues that the platform has already built trust between the participating stakeholders and increased the number of positive Sino-Indian interactions over the YTB (Barua 2018b; A. Barua, personal communication, 20 June 2019).

The Common Threat of Climate Change on Asia's Water Towers

Sino-Indian cooperation over the management of the YTB is even more imperative now, given the predicted impacts of climate change in the river basin. The flows of the YTB and the ferocity of its floods are highly dependent on the melting of Himalayan snow and ice, as well as the arrival of summer monsoon rains (Goswami 1985). According to the IPCC

Fifth Assessment Report, as human activities drive up surface temperatures, the Himalayas could experience between 15% and 78% glacier mass losses by 2100 (Jiménez Cisneros et al. 2014). As glaciers retreat, glacier-fed rivers such as the YTB will first experience an increase in runoff, as more glacial melt swells their flows (Baraer et al. 2012). Climate models also suggest that warming temperatures will increase the frequency of heavy rain events during the monsoon season (Hijioka et al. 2014). Together, an increase in glacial melt from shrinking glaciers combined with an increase in extreme rains during the monsoon season will exacerbate floods for people living in the YTB basin, especially in Assam and Bangladesh (Mohammed et al. 2017). In the long term, as glaciers continue to shrink, the YTB could experience a near 20% decrease in mean upstream water supply between 2046 and 2065 (Immerzeel et al. 2010, 2013), threatening the livelihoods of communities that rely on the YTB flows (Eriksson et al. 2009). Additionally, reduced river runoff jeopardises the success of Chinese and Indian dam-building efforts along the YTB, since hydroelectricity generation depends on river flows.

Given the predicted impacts of climate change on the YTB, it is essential that China and India collaborate to develop a shared adaptation plan for the basin (Ahmad 2019). Additionally, their respective dam-building efforts, if coordinated, could represent a unique opportunity to manage and regulate the flows of the YTB in a way that reduces flood severity throughout the basin (Rampini 2016). The extension of the hydrological data sharing period between China and India could serve as a blueprint for further collaboration in the context of climate change. Indeed, as warmer temperatures alter the flow of the river, it will be key for the two countries to cooperate outside of the regular summer flood season (J. Gupta, personal communication, 27 June 2019).

For now, the only opportunities for Chinese and Indian researchers to meet and discuss the impacts of climate change on the YTB are multilateral platforms such as the Intergovernmental Panel on Climate Change and the Hindu Kush Himalayan Monitoring and Assessment Programme led by the International Centre for Integrated Mountain Development. However, in the context of the Brahmaputra Dialogue, the Indian Institute of Technology in Guwahati (Assam) and Yunnan University have recently signed a MoU to jointly look at the impacts of climate change on water availability and accessibility in the YTB basin. While researchers engaged in this MoU are restricted from sharing their raw discharge data, they will be able to share precipitation, wind speed, and humidity data, and to compare the outputs of their climate models (A. Barua, personal communication, 20 June 2019). This marks the first joint project between a Chinese and Indian institution to focus specifically on the impacts of climate change on the YTB river.

Discussion and Conclusion

The YTB is a massive and complex river system that joins together the fates of China, India, Bhutan, and Bangladesh. Though its flows have remained largely undeveloped, the immense hydropower generation potential of the river has not gone unnoticed, and Asia's two superpowers are now turning their dam-building gaze towards the river. The development of transboundary rivers can strain relations between riparian countries (McCracken and Wolf 2019), and dam-building efforts along the YTB have already caused some friction between China and India, prompting warnings of an imminent 'water war' between the two giants (Chellaney 2011). Moreover, China-Indian relations over the hydropower development of the YBT are complicated by the fact that, as the YTB flows from the Himalayas to the Bay of Bengal, it crosses the territory of Tibet and the disputed eastern Sino-Indian border. As a result, the flows of the YTB are inextricably linked to broader geopolitical conflicts between China and India, such as the issue of Tibet's independence.

Since the 1980s, China and India have made no progress in resolving their border disputes, and both governments have made it clear that they consider these issues, and the question of Tibet, as peripheral to their bilateral relations. However, any meaningful collaboration between China and India over the shared governance of the YTB will require the resolution of these long-standing conflicts, which have led to the militarisation of stretches of the YTB and the securitisation of its flows (R. Bhattacharjee, personal communication, 28 June 2019). At the same time, while government-to-government hydro-diplomacy over the YTB may be unachievable at this time, the relationship between China and India along their shared eastern border and over the YTB is predominantly stable. In fact, Sino-Indian collaboration over the YTB, while still extremely tenuous, has slightly increased since 2013, when China agreed to extend its annual sharing of flow data with India by two weeks. Indeed, China's willingness to share YTB flow data with its downstream riparian neighbours is a crucial step in creating a shared vision for the river, which is imperative given the predicted impacts of climate change in the basin.

Additionally, the relative stability of Sino-Indian relationships in this transboundary region has paved the way for the start of the Brahmaputra Dialogue. As it is, the forum has no executive power to foster the integrated management of the YTB, nor to resolve competing claims between the basin countries. Additionally, its success and continuity are dependent on the maintenance of stable international relationships between China and India, as well as the organisers' capacity to secure funding for these multilateral forums. Still, the forum is the very first platform focused on the YTB that has managed to involve Chinese stakeholders, and the organisers hope that, with time, it could usher in the formation of a transnational and autonomous YTB river basin organisation (A. Barua, personal communication, 20 June 2019).

Overall, while Indian media often portray hydropower development along the YTB as an active area of conflict between China and India (Jiang et. al 2017), relations between the two countries over this shared river are mostly stable. While geopolitical events, such as the succession of the Dalai Lama, could precipitate new developments in the region and in Sino-Indian hydropolitics, it seems more probable that both countries will continue to pursue their own hydro-ambitions without attempting to develop an integrated vision for the river—and with significant implications for Bangladesh, which sits as the lowest riparian country in the basin.

References

Ahmad, O. (2019, February 12). Water dangers loom for South Asia. *The Third Pole*. https://www.thethirdpole.net/2019/02/12/water-dangers-loom-for-south-asia/. Accessed 15 June 2020.

Alam, S., Ali, M. M., & Islam, Z. (2016). Future streamflow of Brahmaputra River basin under synthetic climate change scenarios. *Journal of Hydrologic Engineering, 21*(11), 05016027.

Barua, A. (2018a). Water diplomacy as an approach to regional cooperation in South Asia: A case from the Brahmaputra basin. *Journal of Hydrology, 567*, 60–70.

Barua, A. (2018b, December 27). Best of 2018: Talking to China about the Brahmaputra. *The Third Pole*. https://www.thethirdpole.net/2018/12/27/best-of-2018-talking-to-china-about-the-brahmaputra/. Accessed 15 June 2020.

Bandyopadhyay, J., & Ghosh, N. (2009). Holistic engineering and hydro-diplomacy in the Ganges-Brahmaputra-Meghna basin. *Economic and Political Weekly, 44*(45), 50–60.

Bandyopadhyay, J., Ghosh, N., & Mahanta, C. (2016). *IRBM for Brahmaputra sub-basin water governance, environmental security and human well-being.* New Delhi: Observer Research Foundation.

Baraer, M., Mark, B. G., McKenzie, J. M., Condom, T., Bury, J., Huh, K., et al. (2012). Glacier recession and water resources in Peru's Cordillera Blanca. *Journal of Glaciology, 58*(207), 134–150.

Bhambri, R., & Bolch, T. (2009). Glacier mapping: A review with special reference to the Indian Himalayas. *Progress in Physical Geography, 33*, 672–704.

Bhattacharya, S. (2018, July 23). *China's hydro ambitions and the Brahmaputra.* Institute for Defence Studies and Analyses. https://idsa.in/system/files/backgrounder/b-china-hydro-ambitions-and-the-brahmaputra.pdf. Accessed 15 June 2020.

Blank, J. (2017, September 8). What were China's objectives in the Doklam dispute? *The RAND Blog.* https://www.rand.org/blog/2017/09/what-were-chinas-objectives-in-the-doklam-dispute.html. Accessed 15 June 2020.

Bohmelt, T. (2010). The effectiveness of tracks of diplomacy strategies in third-party interventions. *Journal of Peace Research, 47*(2), 167–178.

Bora, A. K. (2004). Fluvial geomorphology. In V. P. Singh, N. Sharma, & C. S. P. Ojha (Eds.), *The Brahmaputra basin water resources* (pp. 88–112). Dordrecht, The Netherlands: Kluwer Academic.

Burke, J. (2014, April 7). India election begins as north-east's voters kick off world's largest poll. *The Guardian.* https://www.theguardian.com/world/2014/apr/07/india-election-voters-world-largest-poll-modi-rahul-gandhi-bjp-congress. Accessed 15 June 2020.

Central Electric Authority (CEA). (2015). *Review of performance of hydropower stations 2014–15.* New Delhi: Ministry of Power, Government of India. https://www.cea.nic.in/reports/annual/hydroreview/hydro_review-2014.pdf. Accessed 15 June 2020.

Central Tibetan Administration. (2019). *About CTA: Tibet in exile.* https://tibet.net/about-cta/tibet-in-exile/. Accessed 15 June 2020.

Chellaney, B. (2011). *Water: Asia's new battleground.* Washington: Georgetown University Press.

Embassy of the PRC in India. (2019, June 20). The 12th Meeting of China-India Expert Level Mechanism on Trans-Border Rivers was held in India. *Embassy News.* https://in.china-embassy.org/eng/embassy_news/t1674079.htm. Accessed 15 June 2020.

Eriksson, M., Xu, J., Shrestha, A. B., Vaidya, R. A., Nepal, S., & Sanstrom, K. (2009). *The changing Himalayas: Impact of climate change on water resources and livelihoods in the Greater Himalayas.* International Centre for Integrated

Mountain Development. https://lib.icimod.org/record/26471. Accessed 15 June 2020.

Fan, X. (2018, January 17). Opinion: The 'madcap' scheme to divert the Brahmaputra. *China Dialogue*. https://www.chinadialogue.net/blog/10350-The-madcap-scheme-to-divert-the-Brahmaputra/en. Accessed 15 June 2020.

Garver, J. W. (2001). *Protracted contest: Sino-Indian rivalry in the twentieth century*. Seattle: University of Washington Press.

Godbole, S. (2018, May 23). Wuhan Summit: An important signal of intent by India and China. *Brookings Up Front*. https://www.brookings.edu/blog/up-front/2018/05/23/wuhan-summit-an-important-signal-of-intent-by-india-and-china/. Accessed 15 June 2020.

Goswami, D. C. (1985). Brahmaputra River, Assam, India: Physiography, basin denudation, and channel aggradation. *Water Resources Research, 21*(7), 959–978.

He, H. (2011, January 17). A new era for Tibet's rivers. *The Third Pole*. https://www.thethirdpole.net/2011/01/17/a-new-era-for-tibets-rivers/. Accessed 15 June 2020.

Hijioka, Y., Lin, E., Pereira, J. J., Corlett, R. T., Cui, X., Insarov, G. E., et al. (2014). Asia. In V. R. Barros, C. B. Field, D. J. Dokken, M. D. Mastrandrea, K. J. Mach, T. E. Bilir et al. (Eds.), *Climate change 2014: Impacts, adaptation, and vulnerability. Part B: Regional aspects* (pp. 1327–1370). New York: Cambridge University Press.

Hilton, I. (2014). Introduction. In *Brahmaputra: Towards unity* (pp. 4–6). The Third Pole. https://chinadialogue-production.s3.amazonaws.com/uploads/content/file_en/6797/brahmaputra-v13.pdf. Accessed 15 June 2020.

Immerzeel, W. W., Pellicciotti, F., & Bierkens, M. F. P. (2013). Rising river flows throughout the twenty-first century in two Himalayan glacierized watersheds. *Nature Geoscience, 6*, 742–745.

Immerzeel, W. W., van Beek, L. P. H., & Bierkens, M. F. P. (2010). Climate change will affect the Asian water towers. *Science, 328*, 1382–1385.

Jiang, H., Qiang, M., Lin, P., Wen, Q., Xia, B., & An, N. (2017). Framing the Brahmaputra River hydropower development: Different concerns in riparian and international media reporting. *Water Policy, 19*(3), 496–512.

Jiménez Cisneros, B. E., Oki, T., Arnell, N. W., Benito, G., Cogley, J. G., Döll, P., et al. (2014). Freshwater resources. In C. B. Field, V. R. Barros, D. J. Dokken, K. J. Mach, M. D. Mastrandrea, T. E. Bilir et al. (Eds.), *Climate change 2014: Impacts, adaptation, and vulnerability. Part A: Global and sectoral aspects* (pp. 229–269). New York: Cambridge University Press.

Koleski, K. (2017). *The 13th Five-Year Plan. U.S.-China Economic and Security Review Commission*. https://www.uscc.gov/research/13th-five-year-plan. Accessed 15 June 2020.

Krishnan, A. (2013, January 30). China gives go-ahead for three new Brahmaputra dams. *The Hindu*. https://www.thehindu.com/news/international/China-gives-go-ahead-for-three-new-Brahmaputra-dams/article12323702.ece. Accessed 15 June 2020.

Kurian, N. (2013). Uncharted waters: Navigating the downstream debate on China's water policy. In P. J. Das, C. Mahanta, K. J. Joy, S. Paranjape, & S. Vispute (Eds.), *Water conflicts in Northeast India: A compendium of case studies* (pp. 58–64). Pune: Forum for Policy Dialogue on Water Conflicts in India.

Lazerwitz, D. J. (1993). The flow of international water law: The International Law Commission's Law of the non-navigational uses of international watercourses. *Indiana Journal of Global Legal Studies, 1*(1), Article 12.

Liu, J. (2014). The role of civil society in China's anti-dam campaigns. In *Brahmaputra: Towards unity* (pp. 70–73). The Third Pole. https://chinadialogue-production.s3.amazonaws.com/uploads/content/file_en/6797/brahmaputra-v13.pdf. Accessed 15 June 2020.

McCaffrey, S. C. (2008). *Convention on the law of the non-navigational uses of international watercourses: Introductory note*. United Nations Audiovisual Library of International Law. https://legal.un.org/avl/pdf/ha/clnuiw/clnuiw_e.pdf. Accessed 15 June 2020.

McCracken, M., & Wolf, A. (2019). Updating the register of international river basins of the world. *International Journal of Water Resources Development, 35*(5), 732–782.

Ministry of External Affairs (MEA). (2013, October 23). *Memorandum of understanding between the Ministry of Water Resources, the Republic of India and the Ministry of Water Resources, the People's Republic of China on strengthening cooperation on trans-border rivers*. https://www.mea.gov.in/bilateral-documents.htm?dtl/22368. Accessed 15 June 2020.

Ministry of External Affairs (MEA). (2017, May 13). *Official spokesperson's response to a query on participation of India in OBOR/BRI Forum*. https://mea.gov.in/media-briefings.htm?dtl/28463/Official_Spokespersons_response_to_a_query_on_participation_of_India_in_OBORBRI_Forum. Accessed 15 June 2020.

Ministry of Water Resources, Government of India (MOWR). (2017). *India-China cooperation*. https://mowr.gov.in/international-cooperation/bilateral-cooperation-with-neighbouring-countries/india-china-cooperation. Accessed 15 June 2020.

Mohammed, K., Islam, A. S., Islam, G. M. T., Alfieri, L., Bala, S. K., & Khan, M. J. U. (2017). Extreme flows and water availability of the Brahmaputra River under 1.5 and 2 °C global warming scenarios. *Climatic Change, 145*, 159–175.

National Portal Content Management Team (NPCMT). (2011). *International cooperation*. https://archive.india.gov.in/sectors/water_resources/index.php?id=6. Accessed 15 June 2020.

National Remote Sensing Centre (NRSC). (2016). *Flood hazard atlas for Assam state (1998–2015)*. https://asdma.gov.in/pdf/publication/Flood_Hazard_Atlas2016.pdf. Accessed 15 June 2020.

Pahuja, S., & Goswami, D. (2006). *A fluvial geomorphology perspective on the knowledge base of the Brahmaputra*. World Bank. https://archive.org/details/pahuja-and-goswami-2006. Accessed 18 June 2020.

Panda, J. P. (2017). *India-China relations: Politics of resources, identity, and authority in a multipolar world order*. London: Routledge.

Panda, J. P., & Baruah, A. G. (2019). *Foreseeing India-China relations: The 'compromised context' of rapprochement*. East-West Center. https://www.eastwestcenter.org/publications/foreseeing-india-china-relations-the-compromised-context-rapprochement. Accessed 15 June 2020.

Press Trust of India (PTI). (2018, March 28). Indian team in China holds talks on trans-border rivers. *The Hindu*. https://www.thehindu.com/news/international/indian-team-in-china-holds-talks-on-trans-border-rivers/article23370737.ece. Accessed 15 June 2020.

Ray, P. A., Yang, Y. E., Wi, S., Khalil, A., Chatikavanij, V., & Brown, C. (2015). Room for improvement: Hydroclimatic challenges to poverty-reducing development of the Brahmaputra River basin. *Environmental Science and Policy, 54*, 64–80.

Rampini, C. (2016). *Impacts of hydropower development along the Brahmaputra River in Northeast India on the resilience of downstream communities to climate change impacts* (PhD dissertation). UC Santa Cruz.

Singh, P. K. (2018). Resurfacing of divergence in India-China relations. In M. S. Prathiba (Ed.), *China's rising strategic ambitions in Asia* (pp. 87–112). New Delhi: Pentagon Press.

Tandon, A. (2019, September 2). China: Brahmaputra dams won't alter flow. *The Tribune*. https://www.tribuneindia.com/news/archive/nation/china-brahmaputra-dams-won-t-alter-flow-826802. Accessed 15 June 2020.

Vagholikar, N., & Das, P. J. (2010). *Damming Northeast India: Juggernaut of hydropower projects threatens social and environmental security of the region*. Kalpavriksh, Aaranyak, and ActionAid India. https://www.indiawaterportal.org/sites/indiawaterportal.org/files/Damming%20Northeast%20India_%20Kalpavriksh_%20Aaranyak_ActionAid%20India_2010.pdf. Accessed 15 June 2020.

Walker, B. (2016, March 21). China plans more dams and mega infrastructure in Tibet. *The Third Pole*. https://www.thethirdpole.net/2016/03/21/china-plans-more-hydro-projects-and-mega-infrastructure-in-tibet/. Accessed 15 June 2020.

Wasike, S., Okoth, P. G., & Were, E. (2016). The nature of track three diplomacy and its influence on cross-border security relations between Kenya and Somalia. *International Journal of Managerial Studies and Research, 4*(7), 1–12.

Wolf, A. (2003). Conflict and cooperation: Survey of the past and reflection for the future. In F. A. Hassan, M. Reuss, J. Trottier, C. Bernhardt, A. Wolf, J. Mohamed-Katerere et al. (Eds.), *History and future of shared water resources*. New York: UNESCO.

CHAPTER 13

Twenty-First-Century Chinese-African Hydropower Projects in Perspective

Pon Souvannaseng

Introduction

> "Chinese money has to be on the table."—*African hydro-bureaucrat, 2019.*[1]

Sub-Saharan, Central, East, and West African countries in the twenty-first century are seeing a revival of large-scale infrastructure projects, in particular Chinese financed and constructed hydropower dams. Chinese foreign direct investment in Africa grew from 491 million US dollars (USD) to 32.4 billion USD between 2003 and 2014 (UN Habitat et al. 2018). Significantly, Chinese project financing in Africa since the 2008 global financial crisis has continued to expand and is overwhelmingly concentrated in the transport and energy sectors (Garcia-Herrero and Xu 2019;

[1] Interview with African country government hydro-official in April 2019.

P. Souvannaseng (✉)
Bentley University, Waltham, MA, USA
e-mail: psouvannaseng@bentley.edu

J.-F. Rousseau and S. Habich-Sobiegalla (eds.), *The Political Economy of Hydropower in Southwest China and Beyond*, International Political Economy Series, https://doi.org/10.1007/978-3-030-59361-2_13

UNCTAD 2019). More than half of all project finance spending and mergers and acquisitions involving Chinese investors are concentrated in African infrastructure projects, creating large debt-generating flows. While China was a net foreign direct investment importer for much of the end of the twentieth century as it pursued an export-oriented industrialisation strategy, by the start of the new millennium, the Tenth Five Year Plan (2001–2005) was promulgating the Go Out policy directive, which has continued through subsequent Five Year Plans and has become the dominant paradigm for the extension of Chinese capital overseas; this expansion is officially supported and undergirded by coordinated economic statecraft and diplomacy.

The final two decades of the twentieth century, which marked the end of the Cold War, spelled a long period of disastrous and regressive development in Africa. The effects of the 1970s OPEC oil shocks were followed by subsequent debt crises in the developing world and met with strict IMF and World Bank lending policies conditional on structural adjustment austerity measures—dubbed the 'Washington Consensus'. Structural adjustment imposed on African governments resulted in a 'lost decade' of growth and development. Drought, hunger, poverty, conflict, and famine became endemic in the 1980s; in Ethiopia, over a million lives were lost to famine and displacement.

Ethiopia, however, looks remarkably different today, with the motto of its national airline—'the new spirit of Africa'—signalling the country and continent's contemporary turn toward an era of developmentalism. Nowhere is this more evident than in the revival and construction of the 5 billion USD Grand Ethiopian Renaissance Dam (GERD) on the Blue Nile, started under the Meles Zenawi regime in 2011. Once completed, the GERD is expected to increase Ethiopia's total electricity generation capacity by 6000 MW, an increase of 150%. This expansion has transformed Ethiopia into a literal African powerhouse, much to the consternation of its Egyptian and Kenyan neighbours who share the Nile's transboundary river network. Chinese agencies and firms have underwritten over 1.5 billion USD to finance the construction of a power grid in Ethiopia to supply energy to the Chinese financed and constructed high-speed railway network (Champion and Manek 2019). Across the African continent, the 'lost decade(s)' of the twentieth century are quickly transforming into a new millennium of infrastructure-led development underwritten by Chinese sources of finance and players who have usurped the role of traditional Western donors and financial institutions.

From the GERD (6450 MW) and Gibe III (1870 MW) dams in Ethiopia to the Bui (400 MW) and Pwalugu (60 MW) dams in Ghana, and from the Bujagali (250 MW), Isimba (183 MW), and Karuma (600 MW) dams in Uganda to Stiegler's Gorge (1300–1600 MW) in Tanzania, contemporary large hydropower dam building in Africa is on the rise amid poverty and uncertainty over climate change variability. Contemporary African states are experiencing a resurgence of hydropower projects made possible by a new round of foreign financing and significant developments in South–South lending, driven by the rise of Chinese coordinated export promotion and development finance lending. From project finance to construction and operation, Chinese public and private stakeholders have been directly involved in each step of dam development in all of the projects mentioned above. Grand projects from West to East to South Africa are giving the continent's governments a new lease on legitimacy and nation-building—though these projects also come with unintended consequences.

This chapter is organised into three sections. The first discusses different sources of external financing for hydropower projects in Africa across time, spanning the twentieth and twenty-first centuries. This section traces and contextualises changes in comparative regimes of external finance, from US and European dominated financiers to the rise of Chinese project finance in the new millennium. External supply-side and host country demand-side issues around development finance are historically contextualised. The next section discusses how diverse historical hydropower development pathways have converged towards a similar model centred on Chinese coordinated lending. Ethiopia's state-led model, Ghana's bi- and multilateral development finance model, and Uganda's private sector-driven model illustrate different reasons why African dam finance has shifted towards contemporary Chinese financing. The chapter concludes with reflections on the implications of contemporary shifts to Chinese-African hydropower development and an eye towards potential opportunities, perils, and pitfalls.

Background

In this section, I highlight the role of external actors as financiers and the ways in which terms of lending have shifted over time between North-South development actors. This section also traces the widening of the financing landscape to incorporate China as a new development finance lender operating outside of traditional Organisation for Economic Co-operation and Development (OECD) lending norms, with a special focus on how this has impacted developments in African cases. I first discuss the post-World War II Bretton Woods development lending regime, spearheaded in Africa by the US and European development partners, and then highlight why it has given way to new alternative arrangements between China and African governments focusing on supply- and demand-side factors.

Bretton Woods international financial institutions (IFIs) such as the International Monetary Fund and the World Bank, established in the aftermath of World War II, were pivotal in underwriting the construction of large-scale infrastructure in Africa as many newly independent nations embarked on first-time nation-state building projects in the era of decolonisation. Many post-colonial African states at the dawn of political independence were agrarian or organised around primary commodity export for the colonial economy, with little to no domestic savings or resources to draw on to finance and build industries. The role of Bretton Woods IFIs was to provide development lending—at concessional or grant aid terms of lending—to newly independent states in order to jump-start their modernisation drives. It was believed that this large injection of external finance would help newly independent African states leapfrog their way into international industrial ranks. An example of this was the Kariba Dam between Zambia and Zimbabwe, one of the earliest and largest dams in Africa funded by the World Bank in 1956. Subsequent loans established a power station and the building of the Itezhi Tezhi storage dam on the Kafue River, at a cost of 118 million USD lent directly to the Zambia Electricity Supply Corporation Limited, the state-owned power company of Zambia (IBRD 1973).

This high period of state-led development supported by development economics in the 1950s and 1960s coincided with a belief in substituting lacking domestic financial resources, known as the 'savings gap', with concessional aid or finance.[2] These tenets of OECD aid to developing countries, particularly in Africa, established a distinction and separation of principles between developmental donor aid on the one hand and for-profit trade and export promotion on the other. The latter became

[2] The filling of the domestic savings gap in developing countries with external assistance to generate a capital to output ratio was enshrined in the Harrod-Domar model; for a critical take, see Easterly (1999).

enshrined in the OECD Arrangement on Officially Supported Export Credits in 1978, which disallowed OECD countries 'tying' development aid and development finance to other forms of state-to-state trade and business, effectively setting limits and constraints on North-South lending. The OECD Development Assistance Committee (DAC), formed in 1960, also established coordination and norms for aid and development assistance between donor states. The OECD Arrangement on Export Credits and OECD–DAC effectively governed the norms, terms, and purposes of lending, and importantly established clear guidelines for aid and trade as sources of finance between external actors and developing country governments.

OECD governments, as lenders and donors, imposed increasingly strict requirements for access to capital for developing country governments following the 1970s OPEC oil shocks and successive rounds of financial crises, including the end of the gold standard in 1971, the 1997 Asian financial crisis, and the 2008 global financial crisis. Governed by market-reformist ideology and a belief in small government for recipient nations, IFIs and OECD agencies minimised sources of concessional lending and aid for public projects in recipient countries while enhancing and widening space for advanced industrial private sector-led initiatives. This long-term trend in shrinking aid and concessional finance while shifting towards a 'blended finance' approach, which sees a larger role for private sector involvement in public-private infrastructure partnerships, has been consistently promoted since the end of the Cold War by the OECD, G20, and IFIs. It has been renewed in the twenty-first century under a discourse of 'financing the infrastructure gap' by turning 'infrastructure into an asset class' for the financial sector.[3] For African governments, access to concessional and commercial sources of large-scale funds for long-term infrastructure development became nearly impossible within the neoliberal policymaking of the Washington Consensus that prevailed from the 1980s and 1990s into the new millennium.

[3] For more on this, see Mawdsley (2018), and Mawdsley et al. (2018).

Enter China

China's Tenth Five Year Plan (2001–2005) and subsequent Five Year Plans have encouraged tremendous Chinese activity in developing country markets. The promulgation of China's Go Out policy coincided with the dismantlement of China's domestic energy monopoly, known as the State Power Corporation (SPC), and the decentralisation of state entrepreneurialism (Ma and He 2008). As part of the 2002 Scheme of the Reform of the Electricity Industry, SPC was divided into 11 new corporations including five independent power producers (IPP) known as the Big Five, two grid operators, and four auxiliary corporations. The Big Five IPPs—Huaneng Group, Datang Group, Huadian Corporation, Guodian Corporation, and Power Investment Corporation—have become major players in overseas construction in Africa; already-existing major national corporations involved in Chinese domestic energy production such as China International Water and Electric Corporation (a subsidiary of the China Three Gorges Corporation) and China Water Resources and Hydropower Engineering Corporation (Sinohydro) have also expanded into African development. The 2005 Renewable Energy Law aimed at curbing domestic reliance on coal as a core component of China's energy mix also fostered reforms that accelerated the activities of Chinese renewable energy firms domestically and abroad. Three major factors account for this increased economic and diplomatic engagement with the hydro-energy sector in Africa: a policy directive for outward overseas capital expansion set out in the Tenth Five Year Plan; Chinese corporate restructuring with a shift to state capitalism and entrepreneurialism; and a further focus on the renewable energy sector domestically and abroad.

Strong policy directives and coordinated economic statecraft have facilitated Chinese investment and construction projects in Africa. While OECD firms are restricted by competitive market practices which make them averse to the risks of long-term and large-scale infrastructure projects in low-income countries, risk-pursuant Chinese firms have thrived in exactly these less developed spaces. They have been able to utilise development and project finance protocols that reject the constraints of the OECD Arrangement on Export Credits. By effectively tying aid to procurement and construction contracts, plus linking development finance and bilateral trade issues in what is referred to as a 'whole of government' approach, Chinese firms operating abroad have been able to rely on a strong domestic inter-agency apparatus that is willing to use financial and political resources for diplomatic, geopolitical, and commercial ends. Such an apparatus links diplomats, policy banks, and Chinese commercial sector actors together in a coordinated manner. As a result, African host country governments negotiate with Chinese counterparts on a take-it-or-leave-it 'package portfolio' basis. Furthermore, deal-making negotiations between African and Chinese actors are solidified centrally in government-to-government high level meetings which bypass internationally competitive bid processes and competitive project financing procedures.

The Going Out policy, along with its augmentation as the One Belt One Road Initiative in 2013 under Xi Jinping, has been underwritten by China's policy banks in coordination with the Ministry of Finance to direct and suppress its commercial banking sector. In effect, China's hydro-energy infrastructure projects with African counterparts are financed by lending which relies on the Chinese state's control and manipulation of its closed domestic bond market. It has been further undergirded by a state capitalist approach that helped hydropower firms such as goliath PowerChina's SinoHydro to quickly rise in world rankings to become one of the top fifteen international contractors in the world.

African State Building, Legitimacy, and Demand for Resources

For African governments, the lengthy procedures which accompany OECD-financed infrastructure projects (i.e. financial due diligence and environmental and social safeguarding) have made speedy Chinese largesse only too welcome.[4] 'We don't have time for procedure after procedure. We need development. We need it now. We don't have 10, 20 years, we have four' (Interview and participation with African country government hydro-official, April 2019).

For post-independence African state builders, hydropower dams, called 'temples of modernity' by Jawaharlal Nehru, went hand-in-hand with post-colonial nation-building efforts. They symbolised what David Harvey and James Scott have termed high modernist ambitions and were central to forming the foundations for the developing world's

[4] For more on the differences between multilateral development bank, private bank, and Chinese policy bank lending in the hydro-energy industry, see Plummer Braeckman et al. (2020).

industrial aspirations (Harvey 1989; Scott 1998).[5] Ghana's Akosombo Dam (1038 MW), completed in 1965, took pride of place as Kwame Nkrumah's central project and legacy. In neighbouring Egypt, Gamal Abdel Nasser undertook the building of the monumental Aswan High Dam to channel the Nile's waters for agro-economic development purposes.

The post-Cold War introduction of public-private partnership (PPP) structures by OECD lenders and the institutional, regulatory, and bureaucratic reforms which accompanied them opened the pathway for Chinese lenders and firms to move into the financing landscape in Africa. Frustrations mounted in Africa as PPP projects prioritised high tariff rates to pay back external creditors rather than the needs of domestic host country consumers. Starved of international financial lending for long-term infrastructure expansion and squeezed by high electricity tariffs to service foreign debt, many contemporary African countries have experienced power shortages and cuts, prompting strong and popular political impetus for energy projects. Looking to quickly respond to rising energy demand and secure political legitimacy, contemporary African governments see Chinese lending on readily available terms as necessary to secure domestic development goals and political futures.

The next section discusses three African cases with diverse dam financing histories, all of which have converged towards the same outcomes in their willingness to borrow from Chinese lenders to fund energy projects. Following unintended consequences and domestic ramifications from the OECD PPP model, the Ugandan case demonstrates a deliberate pivot towards bilateral Chinese financing. The Ghanaian case illustrates how Western multilateral development bank lending can fall through, providing opportunities for Chinese state business bids. The case also highlights innovations in lending that have allowed the recipient government to receive loans in dyadic forms, which are unconventional to OECD-DAC financing. The Ethiopian case demonstrates how national developmentalist ideology coheres with offshore Chinese lending in ways which conflict with the socio-environmental safeguards and risk frameworks common to traditional OECD donors.

[5] For more on modernisation in Africa, see Cooper (1997).

Comparative Shifts to Chinese-African Hydropower Development

Uganda: From Blended Finance to Coordinated Bilateralism

The Ugandan 250 MW Bujagali power generation project, financed by Wall Street firms and the World Bank, offers an example of OECD dam finance which involves institutional investors and shows how profits are secured across project cycles through a PPP structure. Bujagali achieved financial close in December 2007 and was commissioned in 2011, becoming Uganda's first independent power project (IPP). Financing its 850 million USD total cost involved the largest mobilisation of private finance for a power project in Africa. It was also significant as a landmark case marking the World Bank Group's first return to hydro-financing in Africa after a long period of absence. Bujagali was meant to alleviate severe electricity shortages in Uganda. The IPP was to sell electricity to the Uganda Electricity Transmission Company Limited (UETCL), a parastatal company. However, the financing structure of the project meant that high tariff rates—intended to repay foreign investors for the cost of the project's construction—took priority over setting lower affordable rates for public consumers through UETCL. At nearly ten US cents per kWh, electricity tariffs became unaffordable to a variety of businesses and users, even though the Bujagali Dam increased the country's total energy supply by 45%.

The Bujagali Energy Limited Company (BEL), a special purpose vehicle incorporated in Uganda, was structured to accommodate shared ownership between Industrial Promotion Services Ltd. (IPS), SG Bujagali Holdings Ltd. (SGBHL), and an affiliate of US-based Sithe Global Power LLC. Sithe is indirectly owned by funds managed by the Wall Street private equity firm Blackstone. IPS is the industrial arm of the Aga Khan Fund for Economic Development, the only for-profit vehicle of the Agha Khan Development Network. Commercial lending came from Standard Chartered Bank in the UK and Absa Capital in South Africa. The initial cost of the project was 800 million USD, with the equity portion shared by the private sponsors, IPS, and SG Bujagali Holdings Ltd. On the public side, the government of Uganda committed an in-kind equity contribution of 20 million USD. This effectively made the private-public partnership disproportionately owned and controlled by commercial investors in a build-own-operate-transfer contract arrangement. The

World Bank Group provided 360 million USD of debt funding, including a 130 million USD loan from the International Finance Corporation (IFC) (its for-profit arm), 115 million USD in the form of a partial-risk guarantee to commercial lenders from its donor-funded International Development Association (IDA), and a 115 million USD investment guarantee from the Bank's Multilateral Investment Guarantee Agency (MIGA) to cover the equity position of SGBHL. Loans from the European Investment Bank (EIB), the African Development Bank (AfDB), and a set of European development finance institutions totalling 382 million USD made up the rest. These underwritten assurances from Western donors and various branches of the World Bank worked to incentivise risk-averse Wall Street firms to take part in the project. The IDA partial-risk guarantee covers commercial lenders against debt service default in the event that the government of Uganda fails to meet its payments. Accordingly, the World Bank in effect used the political and bureaucratic enforcement clout of the IDA to ensure against sovereign default. The World Bank IFC loans totalling 130 million USD, MIGA political risk insurance covering the risk of breach of contract for up to 120.3 million USD, and a further 95 million USD on the interest rate swap by lenders brings the World Bank Group's financial risk exposure on the Bujagali HPP project to 374.8 million USD.

In the Bujagali PPP case, the build-own-transfer (BOT) arrangement resulted in a front-loading of high tariffs which made electricity not only unaffordable for Ugandan commercial, industrial, and household consumers, but resulted in the Ugandan government refinancing the project to lower the costs paid to lenders over a longer period. Project debt, which was to be repaid in six years, was extended to 2032. Corporate taxes were also waived to the tune of 28 million USD as part of the refinancing agreement. Refinancing allowed Blackstone to complete the sale of its two-thirds stake in BEL to SN Power AS of Norway for 277 million USD. The Bujagali case illustrates how Wall Street institutional investors like Blackstone profit across the project cycle of dam financing and utilise OECD development finance institutions (DFIs) to ensure guaranteed and timely financial returns while host country governments are constrained by terms of financing and creditor-determined tariffs.

While it took twelve years to finance and build the Bujagali project, the Ugandan government rapidly closed on the 183 MW Isimba Dam in 2013 with bilateral Chinese counterparts. China International Water

& Electric Corporation (CWE) was awarded the engineering, procurement, and construction (EPC) contract worth 568 million USD; the dam was completed just 47 months later, in 2019. A direct concessional loan from the Export–Import Bank of China (China Exim Bank) covered 85% of the cost, with some of it going directly from China Exim Bank to the CWE payroll account, bypassing host country government channels. The Museveni government perceived the BOT-IPP model of Bujagali's development and the high costs associated with compensating Wall Street investors as a debacle. The Ugandan government thus opted for a bilateral deal with a consortium of Chinese public and private actors to build the 600 MW Karuma project in addition to Isimba, as part of a two-part financial package aiming to lower domestic tariffs through further generation. SinoHydro was also awarded the EPC for Karuma as part of the deal.

The Isimba Dam project differs from the Bujagali project significantly and offers a comparable example of what Chin and Gallagher (2019) have called the use of a Chinese 'coordinated credit space' to finance bilateral infrastructure projects. On the surface, the 85/15% division of financing between the 482.5 million USD loan from China Exim Bank and the 15% cash input from the government of Uganda seems relatively straightforward—especially compared to the complex financial and legal engineering involved in the BOT model of financing for Bujagali. The EPC arrangement between the government of Uganda and CWE was a 'turnkey' arrangement in which CWE was only hired as a building contractor with no ongoing role in maintenance and operation. It required the government of Uganda to draw on a consolidated fund established in 2007 to develop generation capacity, advancing payment before China Exim Bank approved the loan (Power Technology 2016).

The risk of a turnkey EPC arrangement is that it incentivises contractors to reduce costs and pocket the difference as profit. This outlook can compromise quality and safety, since the contractor exits the project at the completion of the early construction or 'greenfield' stage, rather than at the later 'brownfield' stage when the dam is operational and running. Uganda's Ministry of Energy and Minerals Development and the Uganda Electricity Generation Company Ltd. were unprepared to technically supervise the projects. For instance, by 2016, cracks and leakages had formed in the Isimba Dam, ultimately causing flooding of the powerhouse and destruction of a coffer dam. An independent panel of experts was formed in April 2016 to guard against inadequate supervision, construction, and quality guarantees at Isimba (Barigaba 2018; UEGCL 2018).

The bilateral financing arrangement of the Isimba and Karuma Dams comes with its own set of risks and drawbacks for the government of Uganda. The two dams are part of a package of numerous other infrastructure projects that will result in a heightened debt ceiling for Uganda, including a 3.2 billion USD standard gauge railway also to be built by Chinese firms. As it attempts to repay the costs of the Isimba project, the government of Uganda must also align payments for Karuma and other infrastructure projects. China Exim Bank's loan requirement comes with a floating interest rate, which means that the interest on the loan may increase substantially over time if market conditions become unfavourable, exposing the government of Uganda to a high insolvency risk. Chinese lenders have safeguarded their loans by securing them with revenue from Uganda's oil reserves, which are also subject to international market price volatility. A joint report by the World Bank and the Ugandan Ministry of Finance Planning and Economic Development warned:

> The government may be tempted to spend in advance some of its future oil revenue through substantial borrowing on the financial markets. This may generate short term benefits through growth of private consumption but would have negative long term consequences as savings are sacrificed (World Bank and Government of Uganda 2015: online).

Ghana: Concessional Finance to Commoditised Loans

Using national natural resources to collateralise dam financing for hydropower projects has enabled Chinese contractors and policy-based lending consortiums to structure and complete financial close in developing countries that would otherwise be considered 'at-risk' of insolvency or plagued by unrated credit status by OECD-based credit agencies. The 400 MW Bui Dam project in Ghana, which reached financial close in 2008 and was completed in 2013, was collateralised with earnings from cocoa exports in a structured deal with China Exim Bank in similar fashion to the oil-backed Isimba project in Uganda.

For Bui, the government of Ghana received two credits from the China Exim Bank: a concessional loan of 270 million USD and a buyer's credit of 292 million USD. These loans were secured against the proceeds of sale of 30,000 tonnes of cocoa a year to China until the project generates enough funds from the sale of electricity to cover the loan repayments. The government of Ghana provided 60 million USD in equity. While this appears simple, the arrangements still took 18 months to complete (Hensengerth 2011).

The pivot to Chinese lending for the Bui Dam marks a shift for the government of Ghana as it moves away from an OECD multilateral development bank model of lending, which had previously financed its landmark first dam, the 1038 MW Akosombo project, and towards Chinese-Ghanaian bilaterally financed and constructed projects. Original funding for the Akosombo project, built in the heyday of the 1960s World Bank era, came in the form of a co-financed 47 million USD International Bank for Reconstruction and Development loan, a direct loan from the US Development Loan Fund of 27 million USD, 10 million USD from the Export–Import Bank of the USA, and a UK export credit guarantee of 14 million USD with a grace period of six years and amortisation period around 25 years. California-based Kaiser Aluminum Corporation, with the support of the US government, received discounted electricity rates from the government of Ghana since Kaiser Aluminum was the main owner of the Volta aluminium smelting plant.[6] Akosombo was thus made possible by a mixture of concessional loan and grant funding from a variety of US-UK and multilateral bank sources.

The state-led development agenda which drove the establishment of the Akosombo Dam differs from the contemporary trend in Chinese financed and constructed projects in Ghana, beginning with the Bui Dam and the further Pwalugu project. Multilateral agencies withdrew support for the Bui project after environmental and social impact concerns were raised about the endangerment of rare animals and livelihood impacts from the flooding of 20% of Bui National Park (Miescher and Tsikata 2009; Hensengerth 2011; Asiama et al. 2017; Yankson et al. 2018). Funding for Bui ultimately came through an unsolicited bid by Chinese funders that was not open to competitive tender processes. The existing Bui Community Development Committee was dismantled and replaced with the Bui Power Authority as a governing institution to accompany Ghana's new external lender. The Pwalugu Dam, a mere 50 MW project, is estimated to cost nearly 993 million USD as part of a multi-project package which includes an irrigation scheme and solar power plant; the

[6] For further treatment and discussion, see Miescher and Tsikata (2009) and Tenkorang et al. (forthcoming).

dam itself is estimated at 366.91 million USD in an EPC arrangement contracted with SinoHydro (GhanaWeb 2020).

Contemporary energy and infrastructure projects in Ghana in which resources serve as collateral, such as the Bui and Pwalugu Dams, have been solidified through high-level political bargaining and issue linkage. Questions have arisen from external and domestic voices about the oversized price tag of projects and their life cycle costs (Awuni 2017). Such developments are geopolitically underwritten by Chinese policy bank financing and mark a departure from traditional OECD multilateral development bank project-based finance. They link projects across sectors, tie procurement, and are not open to competitive tender and bid processes. While China is not the first to collateralise loans with natural resources, it is doing so at a fast pace with broad scale across the globe. That countries like Ghana have taken to resource collateralisation of loans points to their inability to mobilise domestic and other resources or attain alternative lending at concessional value. Ghana's recent move into China's infrastructure finance orbit demonstrates the slippery slope of asymmetric South-South power relations: coordinated bilateralism allows creditors to exert diffuse economic and political influence in borrower countries.

Ethiopia: A Developmental State by Offshore Means

Twenty-first-century Ethiopia, under the leadership of the Ethiopian People's Revolutionary Democratic Front, has developed sub-Saharan Africa's first metro, a sizeable large housing programme, a large army, and the tallest dam in sub-Saharan Africa—a far cry from its 'lost decade' of famine and conflict. Between 2000 and 2014, it borrowed 12 billion USD from China (Rostek-Buetti 2019). The estimated 6000 MW Grand Ethiopian Renaissance Dam (GERD), which may approach 4.5–5 billion USD, was at first financed by Ethiopian domestic resources and strict central government management of coffee exports through its early commodity exchange board (Schäfer 2017). As government coffers emptied, state bureaucrats were enlisted to forego monthly salaries to help pay for the dam (Champion and Manek 2019). Ethiopia's reliance on bilateral Chinese lending to finance its powerhouse ambitions represents a marriage of a particular African developmental state model of domestic resource mobilisation paired with Chinese state and commercial lending. Given constraints to its domestic resources, the Ethiopian Electric Power Company turned to China Gezhouba Group to speed up construction and to access overseas lending (CGTN Africa 2019). Gezhouba is also involved in building Angola's largest dam project, the 4.5 billion USD Caculo Cabaca Dam, which boasts an ambitious capacity of 2.2 GW (GCR Staff 2016). The Industrial and Commercial Bank of China is underwriting the Angolan project as well as a series of further dam projects in Ethiopia along the Omo River, to the consternation of downstream Kenyan authorities.

The 243 m, 1870 MW Gibe III Dam on the Omo River is one of Ethiopia's largest investments and has precipitated grave environmental concerns. Executive decisions taken to assign the EPC contract to Italian infrastructure building firm Salini Impregilo Group under a no-bid process were made with no environmental impact assessment conducted (Woldegebrael 2018). Vicious fighting with the World Bank over environmental safeguards was papered over by domestic rhetoric which framed the project as a national development cause to gain 'victory over Ethiopian enemies'. Pressure from environmental activists dissuaded the AfDB, World Bank, and EIB from offering financial support. Preferential buyer's credit and concessionary loans from China's policy banks allowed the Ethiopian government to finance the 1.75 billion USD Gibe III Dam and the 1.9 billion USD Gibe IV and Warabesa Dams (Dreher et al. 2017). With an expected combined output of 2150 MW, these projects are all being built by SinoHydro. The blending of export credits and development finance in the Chinese-Ethiopian Gibe and GERD deals are considered 'other official flows' of long-term debt—radically differing from the OECD's carefully disentangled official development assistance, private flows, and commercial flows such as foreign direct investment and long- or short-term debt.

The use of Chinese export credit and state-owned commercial financing for the Gibe and GERD projects in Ethiopia highlights how bilateral lending operates unconstrained by international environmental norms and the OECD's Arrangement on Officially Supported Export Credits. China's interest and willingness to lend creates an asymmetric bias for high risk/high reward projects and sectors while dovetailing with the high modernist developmental ideology and ambition of the Ethiopian People's Revolutionary Democratic Front. As government-to-government relations between China and Ethiopia have warmed, Ethiopia's relations with fellow East African nations have grown more tense in the face of local and transnational concerns about the environmental and livelihood impacts of large-scale development (Lavers and

Boamah 2016; Abtew and Dessu 2019; Champion and Manek 2019). The uncompetitive and opaque nature of government-to-government deal making between China and Ethiopia, as elsewhere in Africa, has fostered high-level allegiances. The costs have included rising corruption in host country governments, perilous amounts of debt, and uncertainty regarding how these short-term alliances will play out in the long run as sovereign debt and refinancing come into play.

An Uncertain Future: The Return of High Modernism with Chinese-African Characteristics

In all three cases discussed in this chapter, governments in West and East Africa are deliberately pivoting away from traditional OECD sources of lending to readily available Chinese policy and state-owned commercial financing. African administrations have prioritised lower transaction costs, opacity of business, and speed of construction over life cycle costs and competitive tender and bidding processes to ensure the best value for spending.

In the Ugandan case, the decade-plus experience of the Bujagali project and its unintended consequences created perverse incentives to shift to Chinese bilaterally financed bundled hydropower projects. In the case of Ghana, projects that seemed environmentally and economically unviable have been given a new lease on life through alternative Chinese lending. Unsolicited bidding and innovative resource-collateralised loans have allowed China to gain a foothold in new sectors, projects, and opportunities within and beyond the Ghanaian energy sector. In the case of Ethiopia, a developmentalist regime that had hitherto deployed domestic resources to finance its ambitions has received amplified political and economic support from China in the face of international backlash. This support has allowed the regime to finance its developmental ambitions. All three cases have varied histories of hydro-financing but all have converged on the utilisation and reliance on Chinese lending to help materialise their developmentalist dreams, though with the risk of heavy debt.

Long-term patient capital, once provided by Bretton Woods institutions on concessional and grant basis for state-led development, has been replaced with private sector-driven public-private partnership models. Institutional investors, asset managers, and G20 development agencies in the Global North have not only become more risk-averse in their lending, but also more sensitive to environmental and social concerns from countries with open civil societies that hold them to account. The reluctance of these investors to accommodate African leaders in soft loans and state-led development ideologies has resulted in an about-face by African bureaucrats, who now have a wider landscape of alternative sources of financing. Growing populations and business needs in Africa have created the political impetus for fast solutions and financing.

The rise of alternative South-South lending from Chinese policy banks has coupled with a wider Chinese framework of coordinated export promotion. These financing channels, combined with geostrategic relationship building, have stepped into finance the return of African high modernist ideology writ large across the waterways and ecosystems of the continent. Contemporary Chinese-African hydro-lending is a mixture of commercial and concessional credits that depart radically from OECD norms. South-South Chinese hydropower lending in Africa has entered the energy and finance ministries of Uganda, Ghana, and Ethiopia through opaque and unsolicited channels. Funding arrives with a ready-made system of tied aid, credits, and sector-linked projects in a single portfolio rather than on a project to project basis, making it harder for host governments to negotiate and leverage influence on individual fronts. The engineering, procurement, and construction model of business generates incentives for contracted parties to cut corners and costs and deliver sub-par or dangerous results while exporting technology through tied procurement clauses with little local content inclusion. Without transparent bidding and technical oversight, dam projects underwritten by African host country government guarantees may not be achieving the best value for money over the life cycle costs of projects.

While resource collateralisation as a method for attaining financing is specific and unconventional, it is a double-edged sword between creditor and borrower. Resource-backed loans offer a Faustian opportunity of potential infrastructure-led growth with the risk of potentially crippling long-term debt, asset forfeiture, and the reduction of public spending for social policy goals as budgets are diverted to servicing infrastructure debt. High risk/high reward hydropower projects have pernicious impacts *vis-à-vis* transparency and public accountability on the financial, environmental, and social fronts; these consequences, however, are swept aside by both Chinese and host government protocols in the course of contemporary bilateral megaproject development.

Acknowledgement Funding for parts of this research were provided by the UK Research and Innovation Global Challenges Research Fund and the FutureDAMS research consortia.

References

Abtew, W., & Dessu, S. B. (2019). Financing the Grand Ethiopian Renaissance Dam. In W. Abtew & S. B. Dessu (Eds.), *The Grand Ethopian Renaissance Dam on the Blue Nile* (pp. 161–169). New York: Springer.

Asiama, K., Lengoiboni, M., & van Der Molen, P. (2017). In the land of the dammed: Assessing governance in resettlement of Ghana's Bui Dam project. *Land, 6*(4), 80.

Awuni, F. (2017, April 10). World Bank urges SADA to revise Pwalugu multi-purpose dam project. *CitiFMonline*, http://citifmonline.com/2017/04/world-bank-urges-sada-to-revise-pwalugu-multi-purpose-dam-project/. Accessed 1 June 2020.

Barigaba, J. (2018, April 9). Kampala awaits Isimba power project verdict. *The EastAfrican*. https://www.theeastafrican.co.ke/business/Uganda-awaits-Isimba-power-project-verdict-/2560-4378548-ek4qu5/index.html. Accessed 1 June 2020.

CGTN Africa (2019, February 20). Ethiopia contracts Chinese companies to complete GERD. *CGTN Africa*. https://africa.cgtn.com/2019/02/20/ethiopia-contracts-chinese-companies-to-complete-gerd-construction/. Accessed 1 June 2020.

Champion, M., & Manek, N. (2019, August 1). Death on the Nile haunts Ethiopia's rebirth. *Bloomberg*. https://www.bloomberg.com/graphics/2019-nile-river-ethiopia-dam/. Accessed 1 June 2020.

Chin, G. T., & Gallagher, K. P. (2019). Coordinated credit spaces: The globalization of Chinese development finance. *Development and Change, 50*(1), 245–274.

Cooper, F. (1997). Modernizing bureaucrats, backward Africans, and the development concept. In F. Cooper & R. Packard (Eds.), *International development and the social sciences* (pp. 64–92). Berkeley: University of California Press.

Dreher, A., Fuchs, A., Parks, B. C., Strange, A. M., & Tierney, M. J. (2017). *Aid, China, and growth: Evidence from a new global development finance dataset*. AidData Working Paper #46. Williamsburg, VA: AidData. https://china.aiddata.org/projects/34940?iframe=y. Accessed 1 June 2020.

Easterly, W. (1999). *The ghost of the financing gap: How the Harrod-Domar growth model still haunts development economics*. http://elibrary.worldbank.org/doi/abs/10.1596/1813-9450-1807. Accessed 1 June 2020.

Garcia-Herrero, A, & Xu, J. (2019, July 22). China's investment in Africa: What the data really says, and the implications for Europe. *Bruegel*. https://www.bruegel.org/2019/07/chinas-investment-in-africa-what-the-data-really-says-and-the-implications-for-europe/. Accessed 1 June 2020.

GCR Staff. (2016, December 2). China lends $4.5bn for Angola's Kwanza River mega dam. *Global Construction Review*. http://www.globalconstructionreview.com/news/chinese-lends-45bn-angolas-kwa7nza-rive7r-me7ga/. Accessed 1 June 2020.

GhanaWeb. (2020, February 28). Parliament approves $993 m for Pwalugu Dam project. *GhanaWeb*. https://www.ghanaweb.com/GhanaHomePage/business/Parliament-approves-993m-for-Pwalugu-Dam-project-879844. Accessed 1 June 2020.

Harvey, D. (1989). *The condition of post-modernity: An enquiry into the origins of social change*. Oxford: Blackwell.

Hensengerth, O. (2011). *Interaction of Chinese institutions with host governments in dam construction: The Bui dam in Ghana*. Bonn: German Development Institute.

International Bank for Reconstruction and Development (IBRD) (1973). *Report and recommendation of the President to the Executive Directors on a proposed loan to the Zambia Electricity Supply Corporation Limited with the guarantee of the Republic of Zambia for the Kafue hydro-electric project*. http://documents.worldbank.org/curated/en/392291468170655803/pdf/multi0page.pdf. Accessed 1 June 2020.

Lavers, T., & Boamah, F. (2016). The impact of agricultural investments on state capacity: A comparative analysis of Ethiopia and Ghana. *Geoforum, 72*, 94–103.

Ma, C., & He, L. (2008). From state monopoly to renewable portfolio: Restructuring China's electric utility. *Energy Policy, 36*(5), 1697–1711.

Plummer Braeckman, J., Markkanen, S., Souvannaseng, P. (2020). *Mapping the evolving complexity of large hydropower project finance on low and lower-middle income countires*. FutureDAMS Working Paper 007. Manchester: University of Manchester.

Mawdsley, E. (2018). "From billions to trillions": Financing the SDGs in a world "beyond aid". *Dialogues in Human Geography, 8*(2), 191–195.

Mawdsley, E., Murray, E. W., Overton, J., Scheyvens, R., & Banks, G. (2018). Exporting stimulus and "shared prosperity": Reinventing foreign aid for a retroliberal era. *Development Policy Review, 36*(S1), 25–43.

Miescher, S., & Tsikata, D. (2009). Hydro-power and the promise of modernity and development in Ghana: Comparing the Akosombo and Bui dam projects. *Ghana Studies, 12*(13), 15–53.

Power Technology. (2016). *Isimba hydropower project*. https://www.power-technology.com/projects/isimba-hydropower-project/. Accessed 1 June 2020.

Rostek-Buetti, A. (2019, January 17). China's loan policy under scrutiny. *Deutsche Welle*. https://www.dw.com/en/chinas-loan-policy-under-scrutiny/a-47118123. Accessed 1 June 2020.

Schäfer, F. T. (2017). *Revisiting the agrarian question: Coffee, flowers and Ethiopia's new capitalists*. PhD dissertation. SOAS University of London.

Scott, J. C. (1998). *Seeing like a state: How certain schemes to improve the human condition have failed*. New Haven: Yale University Press.

Tenkorang, E., Enu-Kwesi, F., Bendu, F. S., & Souvannaseng, P. (Forthcoming). *Mapping dam finance projects in Ghana*. FutureDAMS Working Paper. Manchester: University of Manchester.

Uganda Electricity Generation Company Limited (UEGCL). (2018). *Isimba Dam to get new engineers to solve cracks*. https://uegcl.com/corporate-affairs/news/2-isimba-dam-to-get-new-engineers-to-solve-cracks.html. Accessed 1 June 2020.

UN Habitat, et al. (2018). *The state of African cities: The geography of African investment*. UN Habitat and IHS-Erasmus University Rotterdam. https://unhabitat.org/the-state-of-african-cities-2018-the-geography-of-african-investment. Accessed 1 June 2020.

United Nations Conference on Trade and Development (UNCTAD). (2019). *UNCTAD world investment report 2019*. Geneva: UNCTAD. https://unctad.org/en/PublicationsLibrary/wir2019_en.pdf. Accessed 1 June 2020.

Woldegebrael, E. H. (2018). *The materialization of "developmental state" in Ethiopia: Insights from the Gibe III hydroelectric development project regime* (p. 35). Omo Valley: L'Espace Politique.

World Bank, & Government of Uganda. (2015). *Uganda country economic memorandum: Economic diversification and growth in the era of oil and volatility*. Joint Report of the World Bank and the Government of Uganda. http://documents.worldbank.org/curated/en/797831467989562966/pdf/97146-REVISED-V2-PUBLIC-summaryl.pdf. Accessed 1 June 2020.

Yankson, P. W. K., Asiedu, A. B., Owusu, K., Urban, F., & Siciliano, G. (2018). The livelihood challenges of resettled communities of the Bui dam project in Ghana and the role of Chinese dam-builders. *Development Policy Review, 36*(S1), 476–494.

CHAPTER 14

One River and 40+ Dams: The China Factor in the Amazonian Tapajós Waterway

Ricardo Andrade

Introduction

In 2015, during a visit by Premier Li Keqiang to Brasilia, China and Brazil signed 35 bilateral agreements, with planned investments totalling 53 billion US dollars (USD) (DW 2015). This marked an inflection point in trade and investment relations between the countries. In the next few years, Chinese investment in energy and infrastructure in Brazil surged, with Chinese banks offering loans for infrastructure development and large Chinese conglomerates acquiring a variety of assets in the energy sector.

In the first deal after the agreements were signed, the China Three Gorges Corporation (CTG) announced an investment of 3.7 billion USD in hydropower in Brazil, with the company being awarded concessions for the operational Salto, Garibaldi, Jupiá, and Ilha Solteira Dams. In addition, CTG acquired the Santo Antonio do Jari and São Manoel

R. Andrade (✉)
Freie Universität Berlin, Berlin, Germany
e-mail: ricardo.andrade@fu-berlin.de

J.-F. Rousseau and S. Habich-Sobiegalla (eds.), *The Political Economy of Hydropower in Southwest China and Beyond*, International Political Economy Series, https://doi.org/10.1007/978-3-030-59361-2_14

Dams, still under construction, which comprise a portion of the highly controversial Tapajós Waterway (Maloney 2017; Zheng 2017). In 2016, Brazilian oil company Petrobrás contracted a 10 billion USD loan from the China Development Bank (CDB), to be paid in crude oil sent to Chinese companies (Blount 2016). A few months later, CTG acquired the assets of Duke Energy Brazil, including ten hydroelectric plants, in a transaction worth 1.2 billion USD (Zheng 2017).

In 2017 a new 20 billion USD China-Brazil joint fund was announced for infrastructure investment, while State Power Investment Corp (SPIC) bought the São Simão Dam in Goiás (2.2 billion USD) and announced further investment in Brazil (Jornal do Comércio 2017; G1 2019). In the same year, China State Grid (CSG), the largest power utility group in the world, bought CPFL Energia, which operates eight large hydroelectric plants, in a 10.2 billion USD deal. The company plans to invest a further 38 billion USD in Brazil from 2017 to 2021. Notably, it has also been awarded a concession for connecting Belo Monte in the Amazon to the Southwest of the country through two high voltage power transmission lines, set to become the world's longest (2518 km) (Marray 2018; Reuters 2019; Spring 2018).

This series of investments and acquisitions was followed in 2018 by the US-China trade war, which resulted in US President Donald Trump imposing steep tariffs on Chinese imports to the United States and China retaliating by imposing similar tariffs on US imports, including soybeans. US soybean exports to China consequently plummeted from 12.2 billion USD in 2017 to 3.1 billion USD in 2018. As a result, Brazil became the world's leading soybean exporter, further stimulating investment in private ports along the Tapajós River in the Amazon as well as plans for the Tapajós Hydroelectric Complex.

Historically, infrastructure development in Brazil has been carried out by large construction companies (e.g. Odebrecht, Andrade Gutierrez, OAS, Mendes Junior, Camargo Correa, Queiroz Galvão). As revealed by recent high-profile criminal investigations, most of these companies have participated in corruption schemes, usually by siphoning public funds from infrastructure projects to political campaigns (Leahy 2017). Once elected, politicians then later favour the donors in public tenders and approve projects at inflated prices, thus creating a permanent cycle of embezzlement. Despite accusations of political motivations, operation *Lava Jato* (Car Wash), the largest investigation of its kind in Brazilian history, has brought powerful entrepreneurs and politicians to justice, with several still in prison. Most of the companies involved were charged with corruption, had to pay massive fines, had many of their assets confiscated, and were forbidden to provide services to the government. This created a major problem in the Brazilian infrastructure sector, as some of the largest projects underway or planned could not move forward. Chinese finance and infrastructure companies saw this as an opportunity, and the Brazilian government was happy to welcome China's experienced and well-financed infrastructure providers (Leahy et al. 2017).

Until 2015, direct Chinese investment in hydropower projects in Brazil was nearly non-existent. Nevertheless, soybean exports from Brazil to China increased over 5000% from 1998 to 2018 (MDIC 2020), leading major trading and logistics groups to make large investments in private ports along the Tapajós River in the Brazilian Amazon. The main goal driving these investments was to increase the efficiency and reduce the costs of transporting and shipping soybeans to China via the Amazon River, the Atlantic Ocean, and the Panama Canal.

Logistics are a key aspect of the global commodity trade, and lowering the time and cost of transportation are permanent priorities for market operators. Most soy-based products exported from Brazil to China are produced in Central Brazil (Mato Grosso, Goiás) and transported by road to ocean ports (i.e. Santarém and Barcarena in the North and Santos and Vitória in the Southeast). From there, the grain is shipped north along South and Central America and reaches the Pacific through the Panama Canal. Road transport is considered a costly and time-consuming bottleneck: the Mato Grosso State Association of Soy and Corn Producers (APROSOJA) estimates that opening the Tapajós River in the Brazilian Amazon for navigation by large barges would lower transport costs from 227 Brazilian real (BRL) to 60 BRL per tonne, leading to annual savings of 2 billion BRL (470 million USD at November 2019 exchange rates). Dutch Rabobank (2016) forecasts an increase of up to 288% in total export volume via shipping ports in northern Brazil between 2015 and 2025, in contrast with a 2% expected increase for ports in the South that are currently used to ship most of the soy to China. The Brazilian soybean export surge, ongoing since the beginning of the twenty-first century, has benefitted several stakeholders in the soy value chain, with logistics groups reaping major benefits. Eyeing the potential cost reduction that could be achieved by opening the Tapajós River for barge navigation, these companies started buying land and applying for licences to operate private ports in Miritituba, a small town along the Tapajós River that aims to become the main hub for cargo transfer operations in the region. By 2013, logistics groups Rio Turia, Unirios, Reicon, and Brick had announced plans for opening private ports in the city (Aguiar 2017). From 2014 to 2016, five trade and logistics holdings established private ports in the same location: Hidrovias do Brasil, Bertolini, Bunge, Amaggi (which purchased 50% of Bunge's operation), and Cianport. The latter is the only company with a Chinese actor involved; in 2016, Hunan Dakang Pasture Farming Co. Ltd. bought a majority stake in the main subsidiary of the holding company that controls Cianport (Aguiar 2017). Nevertheless, due to the growth in soybean exports to China, the role of the "China factor" in stimulating trade giants and logistics conglomerates to open private ports in the region cannot be denied.

Opening the Tapajós basin for barge navigation drives significant environmental and social costs. In order to tame rapids along the river and increase water depth, the so-called Tapajós Waterway will require the construction of over 40 hydroelectric mainstream and tributary dams and locks (a cascading system known as the Tapajós Hydroelectric Complex). From a social perspective, the project involves the (unconstitutional) forced relocation of a large number of Indigenous tribes who have inhabited these lands for over a thousand years. This is aggravated by the security suspension (Law 4348/1964), a Brazilian legal instrument originating in the military dictatorship, which has been used multiple times in large hydropower projects in order to overturn legal decisions against dam proponents. Security suspension allows chief justices to unilaterally suspend the execution of court rulings against public authorities and their agents in order to avoid damage to the public economy. This chapter will discuss the ways in which the growth in Brazilian soybean exports to China, the construction of private ports in the Amazon, and recent Chinese investment in Brazilian hydropower assets have become intertwined. These factors, along with the questionable legal instrument of security suspension, create a context in which the construction of the Tapajós Hydroelectric Complex seems inevitable, despite enormous social and environmental costs.

Hydropower Development in Brazil and Its Impacts on the Amazon

Up to the late 1960s, wood was still the main energy source in Brazil, supplying 46% of national demand, with oil coming second at 36% (MME 2007). The next two decades saw significant changes in the Brazilian energy matrix: while the economic boom of the early 1970s was largely fuelled by oil, severe crises in oil supply in the mid-1970s and early 1980s led the country to pursue large-scale energy infrastructure programmes based on sugar cane and hydropower to reduce oil dependence. In the late 1970s Brazil started investing heavily in hydroelectric dams: from 1970 to 2012, electricity generation capacity increased over 1300% (IEA 2015) and the share of hydropower in electricity supply grew from 15% to 84.5% (EPE 2013).

In 2007, the Brazilian government published the National Energy Plan 2030 (PNE 2030), a detailed analysis that identifies the main challenges the country might face in meeting energy demand in the next decades. The plan also defines a strategic framework for policy-making and investment in new energy infrastructure. PNE 2030 forecasts a global economic scenario of reasonable growth and stability, but where geopolitical conflicts lead countries to continue grouping in regional blocks (EU, Mercosur, ASEAN, etc.). At the national level, the scenario selected as most probable assumes that key economic problems might have been partially solved by 2030, with a significant reduction in inequality, competitive gains in specific sectors, and medium-high economic productivity. Based on these global and national scenarios, the PNE 2030 estimates a total energy requirement of 402.8 million tonnes of oil equivalent (Mtoe) in Brazil in 2030, with electricity fulfilling 21.18% of total demand (85.3 Mtoe). Considering total electricity generation in 2015, this forecast would require the country to increase its supply by over 100% in the next 15 years. This official forecast of future energy demand in Brazil is strongly contested by local, national, and international NGOs, who argue that the figures are exaggerated and that addressing energy distribution losses would be more rational than increasing production.

The significant investment in hydropower infrastructure carried out by the Brazilian government in recent decades reflects a variety of factors. These span political pressure to address the perceived causes of large-scale energy blackouts, as well as the fact that hydropower is still widely perceived as a clean energy source in comparison with non-renewable alternatives. Systemic corruption also plays a role, as it siphons funds from infrastructure projects for the benefit of electoral campaign financing and makes these projects financially attractive for politicians and large construction companies. This situation is compounded by institutional weaknesses in the legal system, such as political influence in legal decisions about infrastructure projects through the use of the security suspension law.

Despite having currently exploited only about 30% of its hydropower potential, Brazil faces considerable challenges in the planned expansion of the hydroelectricity sector. The Amazon forest and the *Cerrado* (tropical savannah) make up two-thirds of the national territory and hold up to 70% of Brazil's proven and still-untapped hydroelectric potential (MME 2007). These regions host biomes that are under extreme socio-environmental pressure; despite large swathes of land being protected under federal law, they are subject to frequent periods of uncontrolled deforestation, mainly driven by the transformation of primary forests into land for lucrative agriculture and cattle ranching. The Amazon and the Cerrado are also home to many Indigenous communities whose constitutional rights have been historically disrespected in favour of infrastructure development. Abundant water resources, particularly in the Amazon, have led the Brazilian government to plan hundreds of hydropower projects in the region. In 2014, there were 105 operational hydroelectric dams in the Amazon basin (over 70 in the Brazilian Amazon), with 254 under development. In Brazil, these projects have a long record of socio-environmental damage, unchecked political influence on legal decisions, bypassing of constitutional law and binding international treaties, and civil and human rights abuses (Fearnside 2016a, b, c, d).

The Belo Monte Dam in the northern part of the Xingu River illustrates some of the main problems related to hydropower projects in the Amazon. Initial studies to map the hydroelectric potential of the Xingu River started in 1972. In the 40 years that followed from the initial studies to construction and operation, Belo Monte would become the fourth-largest dam in the world (as of 2019), with an overall construction budget skyrocketing from 3.1 billion USD in 2008 to 12.3 billion USD in 2014 (Schmidt 2014). The project has seen conflicts and policy twists that put the best Brazilian soap operas to shame: members of Indigenous tribes pointing machetes at government officials' faces, international celebrities endorsing global campaigns against the project, federal justices voting against constitutional law, top executives of public environmental institutions resigning over political backstabbing, and a small city near the dam becoming the most violent in the country (Instituto Socioambiental 2016; Fearnside 2017a).

From an environmental perspective, the impacts of hydropower dams in tropical regions have been the subject of increasing research (cf. Graham-Rowe 2005). According to Fearnside (2009), the Belo Monte Dam by itself has a small reservoir (440 km^2), and a large installed capacity (11.2 GW). However, the Babaquara Dam, originally planned to be constructed upstream on the Xingu River to regulate water flow and increase power generation at Belo Monte, would flood an area of 6140 km^2. Water levels at Babaquara could rise and fall up to 23 m every year, exposing an area of 3580 km^2 during the low water level months, which would be rapidly covered by vegetation. The decomposition of this vegetation would create a permanent source of methane production. The licence granted for the construction of the Belo Monte project allows only one dam complex to be built in the Xingu basin, but considering the history of bending legal institutions to political pressure in Brazil, one could easily foresee other licences being granted in the future (Fearnside 2017b).

In recent years, international trade of commodities such as soybeans, oil, and iron ore as well as Chinese investments in Brazilian energy infrastructure have played an increasing role in determining the fate of hydropower projects in the Amazon region. In the next section of this chapter, we discuss how the growth in commodities trade between Brazil and China in recent decades and increasing Chinese control of hydropower plants in Brazil have correlated with the opening of private ports along the Tapajós River, influencing governmental plans for the Tapajós Hydroelectric Complex.

The China Factor and the Tapajós Hydropower-Commodities Connection

From 1977 to 2016, land in Brazil used for soybean plantations expanded by 378%, and the country has led the global expansion of soybean planted area since 2000, accounting for 47% of total global growth (Silva et al. 2017). The growth in demand from China and the recent US-China trade conflict, along with productivity increases and favourable policies in Brazil, have established what is known as the "Brazil-China soy trade system"—a transnational, large-scale, highly profitable supply chain of food commodities. This bilateral trade has multiple implications for agriculture, logistics, energy, and infrastructure, and has significant impacts on the well-being of local communities and the preservation of vulnerable ecosystems both in Brazil and China.

The rise in food commodity prices that marked the 2000s and 2010s kept the economies of several developing countries afloat after the 2008 global financial crisis. Food price inflation also contributed to the expansion of the logistics sector in Brazil, leading to increasing interest in major hydroelectric projects in the Amazon. In this section, we will demonstrate how a commodities boom driven by Chinese demand led to a 5000% increase in exports of Brazilian soy to China from 1998 to 2018 and correlated with intensive investment by soybean trade and logistics conglomerates in private ports along the Tapajós River. Mega-projects such as the Tapajós Hydroelectric Complex would allow significant transport cost savings between Mato Grosso (the centre of soy production in the Brazilian Midwest) north to the Amazon River via Miritituba, a small city at the margin of the Tapajós River, where many private ports and cargo transfer stations are currently under development. Increased volumes would then flow through Miritituba, from which soybeans are transported by road to the nearby port of Santarém along the Amazon River; the commodity is then loaded onto larger ships headed to the Atlantic Ocean and finally to China.

As explained above, Brazil-China soy trade growth contributed to Chinese stakeholders becoming much more involved in Brazilian hydropower projects from 2015 onward. Companies such as CSG, CTG, and SPIC have indeed recently acquired significant assets in the Brazilian hydro sector, with Chinese actors currently controlling 10% of energy generation in the country (Polito 2018).

Until 1993, China was a net soybean exporter, but a series of policy changes in the mid-1990s radically altered this scenario. Coping with low yields in 1995 and 1996, Beijing adjusted soybean import policies, decreasing import tariffs from 180 to 3% for imports within allowed quotas. China's soybean imports accordingly increased 86-fold between 1995 and 1999, then amounting to 4.3 million tonnes (Silva et al. 2017).

In 2000, Beijing cancelled all soybean import quota limits and eliminated soy import tariffs, leading to an increase of 113.87% in the export of Brazilian soy to China. In 2001, China joined the World Trade Organisation (WTO), and economic liberalisation together with growing internal

demand sustained multi-year positive outcomes for Brazilian soybean exports. From 2001 to 2004, year-on-year Brazilian soybean sales to China increased, respectively, by 46.4, 74.7, 66.9, and 33.7% (MDIC 2020).

Up to 80% of the soy imported by China is used for animal feed. Significant projected growth in Chinese meat consumption has spurred global commodity giants like Cargill, Bunge, and Amaggi, as well as Chinese banks and state-owned enterprises (SOEs), to invest heavily in the global soybean supply chain. Cargill was the first agribusiness giant to establish a logistics hub in the Tapajós basin: in 2003, the group built a private port in Miritituba, starting operations before final official licensing was issued (Aguiar 2017).

A market contraction in soybean trade in 2005 was followed by several years of further growth in Brazilian soy exports to China (up 35.3% in 2006 and 23.7% in 2007) (MDIC 2020). It is interesting to note that in 2008, despite the global financial crisis and economic meltdown, Chinese imports of Brazilian soybeans increased by 95.37%.

In the meantime, initiatives to explore the hydroelectric potential of the Tapajós basin were also moving forward. In 2009, a consortium formed by Eletronorte, Camargo Correa, and CNEC Engineering commissioned Intertechne Consultants to develop hydraulic inventory studies of the Tapajós River and its tributary, the Jamanxim River (INESC 2020).[1] A few months later, the Brazilian Electricity Regulatory Agency (ANEEL) approved Intertechne's studies, as well as a request by Centrais Elétricas do Norte do Brasil to perform feasibility studies for the Cachoeira dos Patos Dam (528 MW) along the Jamanxim River. The year 2010 was also significant for hydropower development in the Tapajós region: the Brazilian Institute of Environment and Renewable Natural Resources (IBAMA) began the environmental licensing process for five dams with a total reservoir area of 2000 km^2 on the Tapajós River. In the following years, soybean trade continued to grow: while soybean exports to China grew by 9.6% in 2009 and 17.3% in 2010, annual sales peaked at 11.7 billion USD in 2011, a year-on-year surge of nearly 50% (MDIC 2020).

This period was also marked by protest from local Munduruku Indigenous communities against the Tapajós Hydroelectric Complex. Since the

[1] Intertechne's clients also include Odebrecht and Andrade Gutierrez, two of the largest construction companies in Brazil.

eighteenth century, gold has been mined in the Tapajós region where the Munduruku live, and in the 1970s the world's largest gold deposit was discovered in the area. The government usually turns a blind eye to the hundreds of small, unregulated gold mining operations that have been active there for decades. It is not unusual for members of local Indigenous tribes to work in these mines, and the money they earn plays a major role in financing protests against large hydroelectric dams in the Tapajós region. Historical studies show that the Munduruku have only been able to hold on to their territory through active involvement in gold mining (Torres and Branford 2017). In 2007, a young member of a Munduruku tribe was shot dead by a federal police officer during an operation in which 150 armed officers landed by helicopter near the Tapajós River to dismantle an unregulated gold mine. Such violent episodes have led the Munduruku to strongly distrust government institutions together with any initiative related to hydropower expansion. In 2013, as logistics groups started setting up bases in Miritituba, the Munduruku took three researchers hostage who were conducting feasibility studies for hydroelectric dams in the Tapajós region. Local Indigenous communities had not been consulted prior to the studies, as required by the Brazilian constitution and Convention 169 of the International Labour Organisation, of which Brazil is a signatory. The government agreed to stop the studies, and the researchers were released unharmed two days after they were captured. In the same year, the Ministry of the Environment suspended environmental licences for five dams planned in the area due to flawed environmental and social impact assessments.

Meanwhile, Chinese investment in Brazilian energy assets continued to grow: in 2019, CSG announced plans to participate in the privatisation of Brazilian public utility Eletrobrás and pledged new investments of 38 billion USD in Brazil for the next five years. CSG, CTG, and Zhejiang Electric Power Construction have also been involved in initiatives to buy partial or controlling stakes at the Belo Monte Dam (Aguiar 2017), and in the latest BRICS meeting in 2019, Beijing announced investments of 100 billion USD in Brazil, mostly in energy and transport infrastructure (Folhapress 2019).

These initiatives are aligned with Xi Jinping's *Belt and Road Initiative* (BRI), which since 2013 has evolved to become China's main foreign policy initiative and the largest infrastructure development effort in the world. In spite of growing uncertainties regarding the economic feasibility

of many projects and China's capacity to deliver on promised investments (Lam 2016; Hillman 2019), over 125 countries have signed agreements to participate in the BRI (Kuo 2019). Interestingly, notwithstanding the recent large-scale investments by Chinese actors in Brazil and the booming commodity trade, Brazil has not yet signed an official agreement to join the BRI (Green Belt and Road Initiative Center 2020). Nevertheless, it is essential to consider the broader context in order to fully understand the role of China in the development of large-scale hydropower projects in the Amazon, and particularly in the new Tapajós Waterway.

This chapter has so far illustrated the often-convoluted ways in which Sino-Brazilian trade has intertwined with logistics infrastructure development, hydropower projects, and social and environmental conflicts in the Brazilian Amazon. In the next section, we look more specifically into some of the socio-environmental impacts and legal conflicts involved in the Tapajós region.

The Tapajós Waterway, Constitutional Rights, and the Security Suspension Law

The previous sections have unpacked political and economic issues related to Brazil's hydropower expansion, including socio-environmental impacts, policy inconsistencies, and the role of China's economic growth and thirst for commodities in fostering hydropower development in the Brazilian Amazon.

In no other initiative are these factors currently more visible than in the Tapajós Waterway, a government-planned series of up to 43 large hydropower dams located along the Tapajós River and its main tributaries. The many rapids along the Tapajós basin and the presence of multiple *Munduruku* Indigenous communities who have lived in the area for thousands of years have kept the region relatively isolated from previous regional development initiatives. In the last decade, the surge in soybean exports to China—and Beijing's strong interest in controlling energy sources in Brazil and around the world—has changed this scenario. Emboldened by the recent growth in soybean exports, the Brazilian agribusiness and energy sectors have exerted increasing pressure on Brazil's executive and Congress in order to establish the Tapajós Waterway, facilitating logistics and decreasing transport costs for commodities (Fearnside 2015).

The Tapajós Waterway will flood up to 18,700 hectares of land currently inhabited by Munduruku Indigenous communities, and over 950,000 hectares of forest will be cleared (Watts 2017). Initial studies estimate that 32 Munduruku tribes will be affected, and three of them will need to be relocated: Karo Muybu, Sawré Muybu, and Dace Watpu (Sanson 2013; Fearnside 2016b). As with the Belo Monte Dam and most other hydroelectric projects in the Amazon, the drafting of the Environmental Impact Assessment (EIA) for Tapajós-related dams has been marred by multiple problems. The initial documents artificially minimised environmental and social impacts, did not include a formal analysis about impacts on Indigenous peoples, ignored a still-isolated Indigenous tribe from the region, and did not recognise the rights of traditional non-Indigenous communities (*ribeirinhos*) who have also lived in the Tapajós basin for a long time and whose livelihoods depend largely on fisheries and natural resources. The final assessment of the Indigenous component of the Tapajós Waterway EIA concluded—contradicting a formal analysis by FUNAI technicians—that the flooding of Indigenous lands is a historical process that can be mitigated by adequately compensating Indigenous communities who might face losses. This formal document issued by the agency in charge of protecting Indigenous communities goes directly against Article 231 of the Brazilian Constitution and is far from the guidelines defined by International Labour Organisation (ILO) Convention 169 (Fearnside 2016a, b, c, d, e).

The ongoing conflicts between the Munduruku communities and proponents of the Tapajós Waterway include many other inconsistencies. For example, the Brazilian government has been degazetting parts of conservation units that would be affected by the dams, even before projects are evaluated and licensed. Also, demarcation of Indigenous lands in Brazil has come to a halt in recent years, as other powerful ministries involved in the Tapajós Hydroelectric Complex enforce their interests. This became clear in September 2014, when a video was made public showing the then-president of FUNAI, Maria Augusta Assirati, in tears, trying to explain to a group of Munduruku that the paperwork for demarcating their reserve was ready and had been sitting on her desk for a year because other agencies were discussing the proposal. Nine days after the video was made public, Ms. Assirati was demoted from her position due to political pressure from the agribusiness lobby and the Ministry of Mines and Energy (Fearnside 2015).

Political and administrative issues such as the ones described above are compounded by market forces and institutional weaknesses in law enforcement. The booming commodities trade with China has led major market actors to bypass legislation and start commercial operations that provide arguments for dam proponents to justify the hydroelectric and river transport complexes. For example, Cargill's 2003 establishment of a private port in Santarém without any formal licence was perceived by local, state, and federal government bodies as a sign of market interest in the Tapajós Waterway and was followed by the construction of several other private ports in the region. However, due to protests by civil society organisations against the Cargill port, the new ports were set up farther away from the city of Santarém (and closer to Indigenous areas in the middle section of the Tapajós). With the ports established and operational, financing institutions and energy planners foresee economic growth and higher energy demand (Rabobank 2016). This, in turn, fuels powerful private and public actors' interest in further opening the Tapajós River for navigation. The construction of the Tapajós Hydroelectric Complex becomes all the more appealing, reinforcing the arguments of dam proponents.

Another key aspect of the Tapajós Waterway is that completion of *all* the planned dams and locks will be necessary in order to ensure that large barges can navigate the entire Tapajós River and its tributaries. If any of the dams or locks is not built, the final goal of facilitating logistics and decreasing transportation costs for export commodities cannot be achieved. This all-or-nothing condition implies that, if the project moves forward, the entire Tapajós basin ecosystem and all the communities in the region will face the impacts of each specific dam, together with the compounded impacts of all the 40+ dams in the system (Fearnside 2015).

The main legal instrument available for the protection of Brazilian Indigenous communities against power abuse by dam proponents is Article 231 of the 1988 Brazilian Constitution. Paragraph 3 of Article 231 states that the exploration of hydro resources in Indigenous land can only be implemented with the approval of Congress, after local communities are consulted, and requires that these communities share the benefits of such projects. Paragraph 5 prohibits the relocation of Indigenous communities from their land except in the case of natural catastrophes or epidemics that put Indigenous populations at risk, or when national

sovereignty is threatened. The text ensures that if communities are relocated due to any of these special motives, their right to return to the same region is guaranteed once the risk has passed (Fearnside 2016b).

Brazil is also a signatory to the 1989 Convention 169 of the ILO, Article 7.1 of which states that Indigenous communities have the right to choose their own priorities regarding development processes affecting their lives and their land. These communities must also participate in the drafting and implementation of regional and national development plans and programmes that directly affect them. In 2004 the Brazilian government issued Decree 5051, which codified the guidelines approved in ILO Convention 169 into national law (Planalto 2004). Decree 10088, issued in 2019, further details and consolidates these norms (Planalto 2019).

Apart from the legal instruments described above, Brazilian citizens (including Indigenous communities) can also appeal to the Ministério Público (MP, Public Prosecutor's Office), a judicial institution established in 1890 to protect citizens against abuse by the state. In recent decades, and particularly after the 1988 Constitution, the MP has gained considerable autonomy and power.[2] Historically, the MP has been generally perceived by Brazilian society as an efficient and ethical institution, in contrast to the negative perception towards other ministries and public bodies considered highly influenced by politics. Indeed, although legal instruments and institutions aim to protect Indigenous groups' social and constitutional rights, law enforcement and compliance in Brazil is far from ideal. This is especially true in frontier regions such as the Amazon, where the state has a weaker presence and where courts as well as the police are often controlled or heavily influenced by wealthy landowners or corporate interests.

Fearnside (2016a) and Aguiar (2017) highlight how Tapajós Waterway proponents could bypass environmental protection safeguards and the constitutional rights of impacted Indigenous communities. The security suspension law of 1964 (Law 4348) authorises the president of any court of justice to unilaterally suspend the execution of injunctions or rulings against public authorities and their agents in order to avoid damages

[2] The MP was notably responsible for conducting the Lava Jato corruption investigation that resulted in heavy penalties for some of the largest Brazilian construction companies, as well as in the arrest of top executives and politicians. The institution has recently been under public scrutiny for alleged inconsistencies in the legal process that led to the arrest of former President Luis Inácio Lula da Silva.

to the public economy. The law is a legacy of the military dictatorship (1964–1985). In summary, whenever any lawyer or institution (such as the MP) appeals an injunction against any action by public authorities or their agents (such as ministries, public utility companies, licensing agencies, or other dam proponents), the president of the court can use Law 4348 to immediately suspend the injunction. This allows the public action/initiative to proceed unimpeded until the legal case goes through all the possible instances and appeals, up to the final verdict. By 2016, security suspensions had already been utilised twelve times to suspend injunctions against the Tapajós dams (Fearnside 2016a).

The Brazilian judiciary is known for the extremely long delays required for a lawsuit to run its course—it is not uncommon for cases to last 20 or 30 years, and the current backlog of cases numbers close to 100 million. Once a court president suspends an injunction against a certain public initiative, no other legal instruments can be used to block the project from moving forward while the lawsuit runs in parallel in the judiciary. In practice, security suspension allows court presidents to single-handedly overrule injunctions aimed at defending any kind of constitutional right, in order to protect the "public interest". Obviously, any hydroelectric dam is, in principle, of public interest, so Law 4348 allows dam proponents in Brazil to bypass all environmental and social protections, creating a "safety net" for hydroelectric power projects. This presents a major challenge to local communities, NGOs, and other dam opponents: media coverage and public attention usually focus on specific high-profile dams or conflicts, but the way decisions are made is much more important in order to guarantee social and environmental justice in the future (Prudente 2013, 2014; Fearnside 2015).

Conclusion

In this chapter, we have described how Chinese involvement in hydropower projects in the Brazilian Amazon goes beyond financing, acquiring, or operating hydropower dams. By establishing a profitable and large-scale transnational commodity trade system with Brazil, China plays an indirect but decisive role in the definition of national interests and priority infrastructure projects. The decision-making process for hydropower dams in Brazil lacks transparency and policy coherence. Legal instruments such as security suspension further allow dam proponents to

circumvent environmental safeguards and impact Indigenous communities' constitutional rights. In order to ensure democratic policy-making in the hydropower sector, impacted communities, NGOs, activists, and supporting political coalitions should aim, for example, at reforming the legal framework for EIA studies, assuring that they are not paid for by dam proponents (as is often the case), but by independent bodies. Another priority should be reversing the security suspension law, an obsolete legal instrument that has been repeatedly used to neglect the constitutional rights of local communities in the Amazon.

References

Aguiar, D. (2017). *A geopolítica de infraestrutura da China na América do Sul: Um estudo a partir do caso do Tapajós na Amazônia brasileira*. Rio de Janeiro: Actionaid/FASE.

Blount, J. (2016, February 27). China loan to Petrobrás may help most 2016 debt. *Reuters*. https://www.reuters.com/article/us-brazil-petrobras-china/china-loan-to-petrobras-may-help-pay-most-2016-debt-source-idUSKCN0W001T. Accessed 16 June 2020.

Deutsche Welle (DW). (2015, May 19). Brasil e China fecham acordos de US$ 53 bilhões e ajuda à Petrobras. *Deutsche Welle*. https://p.dw.com/p/1FStE. Accessed 16 June 2020.

Empresa de Pesquisa Energética (EPE). (2013). *Balanço energético nacional 2013*. Brasilia: Empresa de Pesquisa Energética.

Fearnside, P. M. (2009). As hidrelétricas de Belo Monte e Altamira (Babaquara) como fontes de gases de efeito estufa. *Novos Cadernos NAEA, 12*(2), 5–56.

Fearnside, P. M. (2015). Amazon dams and waterways: Brazil's Tapajós Basin plans. *Ambio, 44*(5), 426–439.

Fearnside, P. M. (2016a, August 1). A Hidrelétrica De São Luiz Do Tapajós: 3 – O Processo De Licenciamento Na Prática. *Amazônia Real*. https://amazoniareal.com.br/a-hidreletrica-de-sao-luiz-do-tapajos-3-o-processo-de-licenciamento-na-pratica/. Accessed 16 June 2020.

Fearnside, P. M. (2016b, August 8). A Hidrelétrica de São Luiz do Tapajós: 4 – O deslocamento de populações indígenas. *Amazônia Real*. https://amazoniareal.com.br/a-hidreletrica-de-sao-luiz-do-tapajos-4-o-deslocamento-de-populacoes-indigenas/. Accessed 16 June 2020.

Fearnside, P. M. (2016c, August 15). A Hidrelétrica de São Luiz do Tapajós: 5 – O direito de "consulta" dos povos indígenas. *Amazônia Real*. https://amazoniareal.com.br/a-hidreletrica-de-sao-luiz-do-tapajos-5-o-direito-de-consulta-dos-povos-indigenas/. Accessed 16 June 2020.

Fearnside, P. M. (2016d, August 29). A Hidrelétrica de São Luiz do Tapajós: 7 – A barragem e o enfraquecimento da FUNAI. *Amazônia Real*. https://amazoniareal.com.br/a-hidreletrica-de-sao-luiz-do-tapajos-7-a-barragem-e-o-enfraquecimento-da-funai/. Accessed 16 June 2020.

Fearnside, P. M. (2016e, October 3). A Hidrelétrica de São Luiz do Tapajós: 12 – O escândalo do "componente indígena" do EIA. *Amazônia Real*. https://amazoniareal.com.br/a-hidreletrica-de-sao-luiz-do-tapajos-12-o-escandalo-do-componente-indigena-do-eia/. Accessed 16 June 2020.

Fearnside, P. M. (2017a). Belo Monte: Actors and arguments in the struggle over Brazil's most controversial Amazonian dam. *Die Erde: Journal of the Geographical Society of Berlin, 148*(1), 14–26.

Fearnside, P. M. (2017b). Brasil's Belo Monte Dam: Lessons of an Amazonian resource struggle. *Die Erde: Journal of the Geographical Society of Berlin, 148*(2–3), 167–184.

Folhapress. (2019, November 15). China põe US$ 100 bilhões de fundos à disposição do Brasil. *Globo.com*. https://valor.globo.com/brasil/noticia/2019/11/15/china-poe-us-100-bilhoes-de-fundos-estatais-a-disposicao-do-brasil.ghtml. Accessed 16 June 2020.

G1. (2019, September 27). Chineses levam usina de São Simão por R$ 7,18 bilhões. *G1*. https://g1.globo.com/economia/negocios/noticia/chineses-levam-usina-de-sao-simao-por-r-718-bilhoes.ghtml. Accessed 16 June 2020.

Graham-Rowe, D. (2005, February 24). Hydroelectric power's dirty secret revealed. *New Scientist*. https://www.newscientist.com/article/dn7046#.VMPGsS7F_6l. Accessed 16 June 2020.

Green Belt and Road Initiative Center. (2020). *Countries of the Belt and Road Initiative (BRI)*. https://green-bri.org/countries-of-the-belt-and-road-initiative-bri. Accessed 16 June 2020.

Hillman, J. E. (2019, April 3). How big is China's Belt and Road? *Center for Strategic and International Studies*. https://www.csis.org/analysis/how-big-chinas-belt-and-road. Accessed 16 June 2020.

Instituto de Estudos Socioeconômicos (INESC). (2020). *Jamanxim*. https://amazonia.inesc.org.br/banco-de-dados-hidreletricas-na-amazonia/jamanxim/. Accessed 16 June 2020.

Instituto Socioambiental. (2016, December 2). *Belo Monte, um legado de violações*. https://medium.com/hist%C3%B3rias-socioambientais/belo-monte-um-legado-de-viola%C3%A7%C3%B5es-43ea35c973b8#.sfxd48ujg. Accessed 16 June 2020.

International Energy Agency (IEA). (2015). *Data and statistics*. https://www.iea.org/data-and-statistics/data-tables?country=BRAZIL&energy=Electricity&year=1990. Accessed 16 June 2020.

Jornal do Comércio. (2017, May 29). Brasil e China lançam fundo de investimento de US$ 20 bilhões. *Jornal do Comércio*. https://www.jornaldocomercio.com/_conteudo/2017/05/economia/565091-brasil-e-china-lancam-fundo-de-investimento-de-us-20-bilhoes.html. Accessed 16 June 2020.

Kuo, L. (2019, April 24). Belt and Road forum: China's 'project of the century' hits tough times. *The Guardian*. https://www.theguardian.com/world/2019/apr/25/belt-and-road-forum-chinas-project-of-the-century-hits-tough-times. Accessed 16 June 2020.

Lam, W. (2016, April 12). Getting lost in 'One Belt, One Road'. *EJ Insight*. https://www.ejinsight.com/20160412-getting-lost-one-belt-one-road/. Accessed 16 June 2020.

Leahy, J. (2017). The long haul to recovery for Brazil's infrastructure. *Financial Times*. https://www.ft.com/content/9c0595a2-7eb4-11e7-ab01-a13271d1ee9c. Accessed 16 June 2020.

Leahy, J., Schipani, A., & Hornby, L. (2017). Brazil's vulnerability is a big opportunity for Chinese investors. *Financial Times*. https://www.ft.com/content/1d803686-c48e-11e7-b2bb-322b2cb39656. Accessed 16 June 2020.

Maloney, P. (2017, January 3). Duke completes sale of its Latin American assets for $1.2B. *Utility Dive*. https://www.utilitydive.com/news/duke-completes-sale-of-its-latin-american-assets-for-12b/433136/. Accessed 16 June 2020.

Marray, M. (2018, November 20). Chinese bidders spark up for Brazil's electricity transmission auctions. *The Asset*. https://www.theasset.com/belt-road-online/35407/chinese-bidders-spark-up-for-brazils-electricity-transmission-auctions. Accessed 16 June 2020.

Ministry of Development, Industry and Foreign Trade (MDIC). (2020). *Séries Históricas*. https://www.mdic.gov.br/index.php/comercio-exterior/estatisticas-de-comercio-exterior/series-historicas. Accessed 16 June 2020.

Ministry of Mines and Energy (MME). (2007). *Matriz energética nacional 2030*. Brasilia: Ministry of Mines and Energy.

Planalto. (2004, April 19). *Decreto 5.051*. Presidência da República – Casa Civil. https://www.planalto.gov.br/ccivil_03/_ato2004-2006/2004/decreto/d5051.htm. Accessed 16 June 2020.

Planalto. (2019, November 5). *Decreto 10.088*. Presidência da República – Casa Civil. https://www.planalto.gov.br/ccivil_03/_Ato2019-2022/2019/Decreto/D10088.htm#art2. Accessed 16 June 2020.

Polito, R. (2018, October 10). Alvo de Bolsonaro, chineses têm 10% da geração de energia no Brasil. *Valor Economico*. https://valor.globo.com/empresas/noticia/2018/10/10/alvo-de-bolsonaro-chineses-tem-10-da-geracao-de-energia-no-brasil.ghtml. Accessed 16 June 2020.

Prudente, A. S. (2013). O terror jurídico-ditatorial da suspensão de segurança e a proibição do retrocesso no Estado Democrático de Direito. *Revista Magister de Direito Civil e Processual Civil, 10*(55), 108–120. https://www.icjp.pt/

sites/default/files/papers/o_terror_juridico_completo.pdf. Accessed 16 June 2020.

Prudente, A. S. (2014). A suspensão de segurança como instrumento agressor dos tratados internacionais. *Revista Justiça e Cidadania, 165*. https://www.editorajc.com.br/suspensao-seguranca-instrumento-agressor-tratados-internacionais/. Accessed 16 June 2020.

Rasmussen, R., & Ikeda, V. (2016). The impact of port expansion on brazilian soybean output. *Rabobank – RaboResearch*. https://research.rabobank.com/far/en/sectors/grains-oilseeds/build-it-and-they-will-come.html. Accessed 16 June 2020.

Reuters. (2019, October 17). *China's State Grid mulls participating in Eletrobras' privatization in Brazil: CEO*. https://www.reuters.com/article/us-state-grid-ceo-brazil/chinas-state-grid-mulls-participating-in-eletrobras-privatization-in-brazil-ceo-idUSKBN1WW28L. Accessed 16 June 2020.

Sanson, C. (2013, March 29). Projetos para construção de hidrelétricas no rio Tapajós intensificam violações de direitos no Oeste do Pará. *Instituto Humanitas Unisinos*. https://www.ihu.unisinos.br/noticias/518835-projetos-para-construcao-de-hidreletricas-no-rio-tapajos-intensificam-violacoes-de-direitos-no-oeste-do-para. Accessed 16 June 2020.

Schmidt, B. (2014, March 13). Megadams are dismal investments. *Bloomberg*. https://www.bloomberg.com/news/articles/2014-03-13/megadams-are-dismal-investments. Accessed 8 June 2020.

Silva, R., Batistella, M., Dou, Y., Moran, E., Torres, S., & Liu, J. (2017). The Sino-Brazilian telecoupled soybean system and cascading effects for the exporting country. *Land, 6*(3), 53.

Spring, J. (2018, May 30). China's State Grid to invest $38 billion in Brazil power sector. *Reuters*. https://www.reuters.com/article/us-brazil-power-state-grid-corp/chinas-state-grid-to-invest-38-billion-in-brazil-power-sector-idUSKCN1IV2SR. Accessed 16 June 2020.

Torres, M., & Branford, S. (2017, January 11). Ministério Público pede indenização de R$ 10 milhões para vítimas da Operação Eldorado. *The Intercept Brasil*. https://theintercept.com/2017/01/11/ministerio-publico-pede-indenizacao-r-10-milhoes-para-vitimas-da-operacao-eldorado/. Accessed 16 June 2020.

Watts, J. (2017, October 5). Brazil's mega hydro plan foreshadows China's growing impact on the Amazon. *The Guardian*. https://www.theguardian.com/world/2017/oct/05/brazil-amazon-tapajos-hydrovia-scheme. Accessed 16 June 2020.

Zheng, X. (2017, September 2). Brazil unit helps CTG go greener. *China Daily*. https://www.chinadaily.com.cn/business/2017-09/02/content_31450758.htm. Accessed 16 June 2020.

Index

J.-F. Rousseau and S. Habich-Sobiegalla (eds.), *The Political Economy of Hydropower in Southwest China and Beyond*, International Political Economy Series, https://doi.org/10.1007/978-3-030-59361-2

The manufacturer's authorised representative in the EU is Springer Nature Customer Service Centre GmbH, Europaplatz 3, 69115 Heidelberg, Germany. If you have any concerns regarding our products, please contact ProductSafety@springernature.com

Printed and bound by CPI Group (UK) Ltd, Croydon, CR0 4YY
30/09/2026
02225702-0017